NORTH CAROLINA
STATE BOARD OF COMMUNITY COLLEGES
LIBRARIES
ASHEVILLE-BUNCOMBE TECHNICAL COMMUNITY COLLEGE

DISCARDED

JUN 1 6 2025

MIXED PLASTICS RECYCLING TECHNOLOGY

MIXED PLASTICS RECYCLING TECHNOLOGY

by

Bruce A. Hegberg
Gary R. Brenniman
William H. Hallenbeck

University of Illinois
Center for Solid Waste Management and Research
Chicago, Illinois

NOYES DATA CORPORATION
Park Ridge, New Jersey, U.S.A.

Copyright © 1992 by Noyes Data Corporation
Library of Congress Catalog Card Number: 91-44920
ISBN: 0-8155-1297-X
Printed in the United States

Published in the United States of America by
Noyes Data Corporation
Mill Road, Park Ridge, New Jersey 07656

10 9 8 7 6 5 4 3 2 1

Library of Congress Cataloging-in-Publication Data

Hegberg, Bruce A.
    Mixed plastics recycling technology / by Bruce A. Hegberg, Gary R.
Brenniman, William H. Hallenbeck.
      p.   cm.
    Includes bibliographical references (p.) and index.
    ISBN 0-8155-1297-X
    1. Plastic scrap--United States--Recycling. 2. Plastics industry
and trade--United States. I. Brenniman, Gary R. II. Hallenbeck,
William H. III. Title.
TD798.H44   1992
363.72'88--dc20   91-44920
                                                                    CIP

Foreword

This book presents an overview of mixed plastics recycling technology. In addition it characterizes mixed plastics wastes, and describes collection methods, costs, and markets for reprocessed plastics products. While these studies were done for the State of Illinois, with the current national concern about recycling, the information presented will be of interest to anyone involved in municipal recycling, and the subsequent processing of post-consumer mixed plastics.

The term "mixed plastics" implies a mixture of plastic resins, or a mixture of package/product types, which may or may not be the same plastic type or color category. The term also includes products which may be the same resin type, but which have been fabricated using differing manufacturing techniques.

1989 data indicate that U.S. plastic resin production totalled 58.2 billion pounds (of which the packaging sector accounted for 14 billion pounds). About 29 billion pounds were disposed of as municipal solid waste (MSW), yet only an estimated 340 to 400 million pounds of plastics were recovered or recycled in any way that year. It becomes evident immediately that plastics recycling holds great potential as an industry coming of age.

The book is presented in two parts. Part I identifies the compositions of plastics in MSW and in recycling programs, the post-consumer plastics contributions to recycling programs, the cost of plastics collection and some of the end uses for reprocessed post-consumer plastics. Attention is given to curbside collection of recyclables because of its high recovery rate (60-90%) in comparison to other recycling methods (10-30%).

Part II discusses technologies which have been developed for the separation and processing of mixed plastic wastes. Broad scale recycling of post-consumer plastic waste is technically difficult because of the variety of plastic resins which exist and the difficulty of sorting them. While further work in processing and separating waste plastics is necessary for widespread plastics recycling, there are methods to utilize mixed plastic waste, and methods to clean and separate some types of plastics. The latter is primarily an emerging field of research in recycling technologies. Recent advances in automated sorting, by plastic type and by color, hold promise for future profit and fewer problems for this industry.

The information in the book is from the following documents:

Post-Consumer Mixed Plastics Recycling—Characterization, Collection, Costs and Markets, prepared by Bruce A. Hegberg, William H. Hallenbeck, and Gary R. Brenniman of the University of Illinois Center for Solid Waste Management and Research for the Illinois Department of Energy and Natural Resources, Office of Solid Waste and Renewable Resources, January 1991.

Technologies for Recycling Post-Consumer Mixed Plastics—Plastic Lumber Production, Emerging Separation Technologies, Waste Plastic Handlers and Equipment Manufacturers, prepared by Bruce A. Hegberg, Gary R. Brenniman and William H. Hallenbeck of the University of Illinois Center for Solid Waste Management and Research for the Illinois Department of Energy and Natural Resources, Office of Solid Waste and Renewable Resources, March 1991.

The table of contents is organized in such a way as to serve as a subject index and provides easy access to the information in the book.

Advanced composition and production methods developed by Noyes Data Corporation are employed to bring this durably bound book to you in a minimum of time. Special techniques are used to close the gap between "manuscript" and "completed book." In order to keep the price of the book to a reasonable level, it has been partially reproduced by photo-offset directly from the original reports and the cost saving passed on to the reader. Due to this method of publishing, certain portions of the book may be less legible than desired.

NOTICE

The materials in this book were prepared as accounts of work sponsored by the Illinois Department of Energy and Natural Resources. On this basis the Publisher assumes no responsibility nor liability for errors or any consequences arising from the use of the information contained herein.

Mention of trade names or commercial products does not constitute endorsement or recommendation for use by the Agency or the Publisher. Final determination of the suitability of any information or product for use contemplated by any user, and the manner of that use, is the sole responsibility of the user. The book is intended for information purposes only. The reader is warned that caution must always be exercised when dealing with plastic wastes, recycling materials or equipment which might be potentially hazardous, and expert advice should be obtained before implementation of recycling procedures.

All information pertaining to law and regulations is provided for background only. The reader must contact the appropriate legal sources and regulatory authorities for up-to-date regulatory requirements, and their interpretation and implementation.

The book is sold with the understanding that the Publisher is not engaged in rendering legal, engineering, or other professional service. If advice or other expert assistance is required, the service of a competent professional should be sought.

Contents and Subject Index

PART I
MIXED PLASTICS RECYCLING—CHARACTERIZATION, COLLECTION, COSTS, MARKETS

SUMMARY 3

1. INTRODUCTION 7
   1.1 National Production and Recycling Levels of Plastics 11
   1.2 Plastics in Municipal Solid Waste 15
   1.3 Mixed Plastics in Post-Consumer Recycling 17

2. CHARACTERIZATION, GENERATION AND COLLECTION
   OF PLASTICS 19
   2.1 Field Assessment of Plastic Types in Municipal Solid
       Waste 19
   2.2 Mixed Plastics in Recycling Programs 23
          New Jersey 23
          Hennepin County, Minnesota 26
          Walnut Creek, California 29
          Akron, Ohio 29
          Ontario, Canada 30
          Seattle, Washington 30
          PS Collection Data 30
   2.3 Per Capita Generation 31
   2.4 Commercial and Food Sector Sources of Waste Plastic 39
   2.5 Post-Consumer Plastic Weights 42
   2.6 Summary 42

viii Contents and Subject Index

3. PLASTICS RECYCLING PROGRAMS 44
   3.1 Curbside Collection of Plastics in Illinois 44
   3.2 Film/Rigid Plastics Recycling 44

4. RECYCLING COSTS 51
   4.1 Recycling Program Variables 51
   4.2 Recycling Costs 52
       Madison, Wisconsin Recycling Cost Estimate 52
       Minneapolis, Minnesota Plastic Recycling Cost
         Estimate 55
       Other Recycling Cost Estimates 58
   4.3 Collection Times 58
   4.4 Recycling Truck Costs and Truck Collection Methods
       for Plastics 61
   4.5 Process Cost 62
   4.6 Cost Estimate Computer Programs 63
       Eastman Chemical 63
       Wasteplan 65
       Least-Cost Scheduling 65

5. MARKETS AND PACKAGING CHANGES FOR
   RECYCLED PLASTICS 70
   5.1 Recycled Resin Demand 70
   5.2 Packaging Changes to Increase Recycle Rates 72
   5.3 Markets in Primary Recycling 74
   5.4 Markets in Secondary Recycling 75

APPENDIX A: RECYCLING VEHICLE EQUIPMENT
MANUFACTURERS 78

APPENDIX B: GLOSSARY 80

REFERENCES 83

PART II
RECYCLING TECHNOLOGY

SUMMARY 89

1. INTRODUCTION 94
   1.1 Plastics in Municipal Solid Waste 94
   1.2 Plastic Resin Production and Product Manufacture 96
       Resin Manufacture 96
       Additives 100

Product Manufacture 100
　　　Extrusion Molding 102
　　　Blow Molding 103
　　　Injection Molding 106
　　　Film Manufacture 107
　　　Melt Flow Indexes 107

2. **MANUFACTURE OF PLASTIC LUMBER USING MIXED PLASTICS** 110
　2.1　Plastic Wood Producers 111
　2.2　Plastic Wood Production 111
　2.3　General Guidelines for Plastic Lumber Manufacturing 115
　2.4　Products from Mixed Plastic Lumber 116
　2.5　Enhancement of Plastic Wood Properties 116
　2.6　Wood Fiber–Resin Composite Lumber 119
　2.7　Future of Mixed Plastic Lumber 119

3. **EMERGING METHODS FOR PROCESSING AND SEPARATION OF PLASTICS** 120
　3.1　Optical Color Sorting of Glass and PET Containers 120
　3.2　Separation of PVC Bottles from Other Plastic Containers ... 126
　3.3　Separation of HDPE Base Cups from PET Beverage Bottles 130
　3.4　Separation Using Selective Dissolution 134
　3.5　Separation Using Soluble Acrylic Polymers 141
　3.6　Initial Activities in Polyurethane Recycling 142
　3.7　Initial Activities in Automotive Plastics Recycling 142
　　　　Identification of Plastic Auto Parts 142
　　　　Solvent Dissolution of Plastic Auto Shredder Residue ... 143
　　　　Pyrolyzing of Auto Thermosets 144
　3.8　Sources of Plastic Recycling Information and Plastic Recycling Systems 144

4. **BUYERS AND SPECIFICATIONS FOR WASTE PLASTICS** 145
　4.1　Buyers of Waste Plastic 145
　4.2　Specifications for Waste Plastic 147

APPENDIX A: PLASTIC SCRAP HANDLERS AND BROKERS 151

APPENDIX B: SOURCES OF INFORMATION ON PLASTICS RECYCLING 185

APPENDIX C: MANUFACTURERS OF PLASTIC RECYCLING EQUIPMENT 191

APPENDIX D: GLOSSARY 200

REFERENCES 205

Part I

Mixed Plastics Recycling—Characterization, Collection, Costs, Markets

The information in Part I is from *Post-Consumer Mixed Plastics Recycling—Characterization, Collection, Costs and Markets,* prepared by Bruce A. Hegberg, William H. Hallenbeck, and Gary R. Brenniman of the University of Illinois Center for Solid Waste Management and Research for the Illinois Department of Energy and Natural Resources, Office of Solid Waste and Renewable Resources, January 1991.

Acknowledgments

This public service report is a result of the concern of the Illinois Governor, State Legislature, and the Public for the magnitude of the solid waste problem in Illinois. The concern led to the passage of the Illinois Solid Waste Management Act of 1986. One result of this Act was the creation of the University of Illinois Center for Solid Waste Management and Research. The Office of Technology Transfer (OTT) is part of this Center. One of OTT's means of transferring technology is the publication of pubic service reports which contain discussions of important topics in solid waste management.

Funding for this public service report was provided by the Illinois Department of Energy and Natural Resources (IDENR), Office of Solid Waste and Renewable Resources. Additionally, OTT would like to acknowledge the review provided by IDENR.

Summary

Recycling of plastic discards is one method of reducing municipal solid waste. They are beginning to join glass, steel, aluminum and paper as waste stream components that have been accepted into recycling programs across the country. It is difficult, however, to expand post-consumer plastics recycling beyond the easily recognized milk jugs and soda bottles because of the variety of plastic wastes, the difficulty of sorting plastic resins, the low density of post-consumer plastics wastes and the limited history of plastics recycling. However, in order to expand the recovery and recycling of plastics and decrease the amount of waste disposed in landfills, it will be necessary to overcome these difficulties. Because of its heterogeneous nature and the amount of contaminants present, separation of post-consumer mixed plastic waste is the most difficult. The term "mixed plastics," a mixture of plastic resins or a mixture of package/product types which may or may not be the same plastic type or color category, has been used to describe broad scale processing of post-consumer plastic waste. Mixed plastics also includes products which may be the same resin type but which have been fabricated using the differing manufacturing techniques. The purpose of this report is to identify the compositions of plastics in municipal solid waste (MSW) and in recycling programs, the post-consumer plastics contributions to recycling programs, the cost of plastics collection and some of the end uses for reprocessed post-consumer plastics. Attention is given to curbside collection of recyclables because of its high recovery rate (60-90%) in comparison to other recycling methods (10-30%).

The 1989 production of plastic resins in the U.S. totaled 58.2 billion pounds. Almost all of the annual production (92%) was consumed in the U.S. Eight resin types make up 83% of the annual domestic demand: low density polyethylene (LDPE), 9.7 billion lbs; polyvinyl chloride (PVC), 7.6 billion lbs; high density polyethylene (HDPE), 7.4 billion lbs; polypropylene (PP), 6.2 billion lbs; polystyrene (PS), 5.0 billion lbs; polyurethane (PUR), 3.2 billion lbs; phenolic, 3.1 billion lbs; and polyethylene terephthalate (PET or PETE), 1.9 billion lbs. The packaging sector is the leading consumer of plastic resins at about 14 billion pounds annually.

A relatively small amount of plastic is recycled on an annual basis in comparison to the production levels of plastic resins or the amount disposed in MSW landfills. It has been estimated that 340-400 million pounds of plastics were recovered or recycled in some fashion in 1989. Approximately 29 billion pounds were disposed in MSW. About half of the recycled plastic came from the recycling of PET beverage bottles (including the HDPE base cup on such bottles), and most of the remainder came from HDPE bottles, PET x-ray film and PP car battery cases. The amount of plastics recycled in comparison to the amount disposed is 1.3%, and in comparison to the annual production level of plastics in the U.S. is 0.6%. A review of the 15 primary resins produced in the U.S. shows that five of the above mentioned resins are disposed of primarily through MSW, and that the remaining plastics are generally destined for non-MSW disposal, i.e.

imports/exports of plastics, recycling, construction/demolition debris or incineration. PVC is the only major resin which is not primarily disposed through residential, commercial or institutional MSW.

To help increase the recycling of plastics in Illinois, the state passed legislation requiring the labeling of six plastic types on all plastic bottles with a capacity of 16 fluid ounces or more and on all other rigid plastic containers with a capacity of 8 fluid ounces or more (PETE - 1, HDPE - 2, PVC - 3, LDPE - 4, PP - 5, PS - 6, and all others - 7). Many manufacturers are now voluntarily labeling their packaging with the appropriate number, even though not required by law. A law has also been passed which requires all counties to develop plans which will achieve 25% recycling. Increasing the recycling of these six primary plastics is a logical next step in satisfying the state law to achieve 25% recycling.

There is a wide variation in the types of plastics currently collected in curbside recycling programs. While some communities collect clear HDPE beverage bottles and/or PET beverage bottles, others have moved beyond this to additionally collect colored HDPE bottles (typically household chemical bottles), any type of plastic bottle, any type of rigid plastic container (RPC), or any plastic with the previously mentioned 1 through 7 numbering system. As part of pilot programs, some communities have started collection of non-bottle packaging such as foam PS and LDPE six-pack rings.

Collection of any type of cleaned plastic, including films, has been conducted in pilot programs and continues in municipalities near "plastic lumber" manufacturers. Towns around Toronto, Canada, on Long Island, New York, in central Michigan, and in northwestern Iowa are areas where such mixed plastic collection is being done. Previous attempts at such extensive mixed plastic collection have not always been successful due to a high reject rate of unacceptable materials (such as rubber hoses and household medical waste), and due to food and container content contamination. The result is a high cost for manual sorting. Resident education, including cleaning and proper preparation of recyclable plastic, was a key element often cited as a method to resolve these problems.

In 1989 it was estimated that 9 million U.S. households were part of curbside recycling programs and that 20% collected some type of plastic. Twenty percent of U.S. households are projected to have curbside collection by 1992 with an annual plastics collection of 334 million pounds. In Illinois by the end of 1990, 600,000 households were expected to have curbside recycling. It is currently estimated that 43 municipalities in Illinois collect some type of plastic, affecting a total of 221,000 residences. Of these, approximately 40% collect some type of mixed plastic. None of the curbside programs in Illinois collect LDPE films and only a few collect RPC or any type of plastic bottle.

A review of plastic in MSW indicates that plastics comprise 6 to 10% of MSW by weight and that the largest constituent is LDPE (film and rigid plastic containers) followed by HDPE. A waste composition analysis for the city of Chicago determined 9.4% by weight of its MSW was plastic. Plastics generally occupy 20% by volume of landfill discards.

Although LDPE is the primary constituent of plastic in MSW, little film plastics (LDPE) recycling is done. Little data is available on post-consumer collection of LDPE. When LDPE is collected in recycling, it can be expected to comprise 25% by weight of the plastics collected.

There are large variations in the composition of plastics collected and the amount collected per household on a weight basis. For a municipality considering collection of mixed plastics, it is best to conduct a pilot program in a test area(s) whereby composition and generation rates can be developed. When measurement of plastic recyclables is based on the entire number of residences in a collection area and includes participants as well as non-participants, the unit "lb/collection area household/year" is used. Generally, collection of RPC will result in 30 lb/collection area household/year (lb/cahh/yr), while collection of plastic bottles will result in 20 lb/cahh/yr, as a minimum. Collecting only clear HDPE (milk) bottles will obtain 5-7 lb/cahh/yr; adding colored HDPE bottles will add another 1-4 lb/cahh/yr; collecting PET bottles will obtain another 2-5 lb/cahh/yr; and collecting PS can obtain 5 lb/cahh/yr. Examining household generation in this fashion will provide an indication of the overall effectiveness of a curbside recycling program with regards to participation and recycling education.

Actual recycling rates for plastics based on only participants which set plastics out at curbside are significantly higher. When measurement of plastic recyclables is based only on the residences in a collection area which participated in plastics recycling, the unit "lb/participant household/year" is used. This type of measurement more appropriate for comparison among curbside recycling programs than measurements which factor in non-participants. Contribution of just HDPE bottles has been shown as high as 35 lb/participant household/year (lb/phh/yr), and PET has been shown at 23 lb/phh/yr. It is further estimated that 6 lb/capita/yr LDPE, 3.5 lb/capita/yr PP and 1.5 lb/capita/yr PVC can be collected in residential recycling. These values can provide an estimate of the actual potential for recycling in an area if all or most residences were to participate in a curbside recycling program. Collection of any type plastic bottle will result in a 75 to 200% by weight increase over the amount collected if only PET and clear HDPE beverage bottles were in a plastics recycling program.

Contamination of non-specified plastic and non-plastic contaminants are usually present in collected plastics. Non-specified plastics range from as little as 1% for relatively simple plastics collection to as much as 10 to 20% for plastic bottle or colored HDPE bottle recycling. Non-plastic contaminants will comprise anywhere from 1 to 10% contamination by weight. Education is the most important way to keep contamination levels down.

The cost associated with plastics recycling depends on many variables such as equipment used, collection methods, collection frequency and material collected. Automation of collection and processing methods have been and continue to be a barrier in reducing the cost of plastics recycling. Generally, the cost of weekly curbside recycling ranges from $12 to $30/household served/year. Collecting weekly rather than biweekly will add 10 to 35% to the cost of curbside collection. There is little information regarding the incremental cost associated with collecting types of plastics. One series of pilot projects determined that adding collection of clear HDPE and PET beverage bottles to a 50,000 household program would cost $0.70 to $1.40/household served/year (using curbside sorting). The cost of adding collection of all plastic bottles, RPC, or clear/colored HDPE and PET bottles would be $3.30 to $4.30/household served/year (also using curbside sorting). None of these values include the cost savings associated with diversion credits or revenue sales, and they also do not include the cost of processing. The cost of baling sorted plastic is estimated at 3 to 4¢/lb. The cost of processing commingled plastic at a sorting and baling/grinding facility is estimated at 10 to 12¢/lb.

Computer programs which assist in solid waste planning and recycling costs are available from the IDENR and from the National Association for Plastic Container Recovery. Another method of solid waste cost planning called "least cost scheduling" uses linear programming to minimize the costs associated with recycling, waste disposal and landfilling. It can also be utilized for cost estimating recycling programs.

The market outlook for recycled plastics is strong. The recycled resin demand for the six previously mentioned plastics (HDPE, LDPE, PET, PP, PS and PVC) by the end of 1993 are expected to be 3.5 times that of 1990 levels. The price of sorted, baled plastic is currently in the range of 7 to 12¢/lb and the price of sorted, cleaned and flaked plastic is 20 to 30¢/lb. Pelletized recycled resin ranges from 20 to 45¢/lb.

To increase the amount of collected plastic, reduce the cost associated with waste plastic processing, and increase the market prices for recycle resin, changes in packaging design are taking place. This includes using reusable packages, using single material packaging wherever possible, using materials that are either easily separated or compatible if a single material cannot be used, using recycled materials where possible, and eliminating toxic constituents from packaging.

1. Introduction

Recycling of plastic discards is one method of reducing municipal solid waste. They are beginning to join glass, steel, aluminum and paper as waste stream components that have been accepted into recycling programs across the country. It is difficult, however, to expand post-consumer plastics recycling beyond the easily recognized milk jugs and soda bottles for technical, economic and social reasons [U.S. EPA, 1990a]:

- *The variety of plastic wastes* Plastics in municipal solid waste (MSW) are a very heterogeneous collection of materials that encompass not only a broad range of types made from a single resin, but also an increasing number of items that include a blend of resins, either mechanically or chemically bonded together. The varieties are made additionally diverse through the use of plastic additives to yield specific product qualities.

- *The difficulty of sorting plastic resins* It is technically difficult to obtain relatively pure resins from mixed plastics collected for recycling. Commercially demonstrated separation technologies are almost exclusively limited to processes that separate polyethylene terephthalate (PET) and high density polyethylene (HDPE).

- *Low density of post-consumer plastics wastes* Plastics occupy a high volume/weight ratio compared to other recyclable constituents in MSW, and this adversely affects the practicality and economics of plastics collection in a recycling program. Landfill diversion rates are measured on a weight basis and the weight contribution of plastics to MSW is relatively small (even though the landfill volume occupied by plastics is large). The large volume occupied by plastics in a recycling truck can displace the ability to collect other "heavier" recyclables.

- *Limited history of plastics recycling* For many plastics recycling alternatives, only limited data exist from which to extrapolate costs, participation rates, technological or institutional barriers, and other factors which affect long-term viability.

However, in order to expand the recovery and recycling of plastics and decrease the amount of waste disposed in landfills, it will be necessary to overcome these difficulties. Because of its heterogeneous nature and the amount of contaminants present, separation of post-consumer mixed plastic waste is the most difficult. Waste plastics from industrial operations are cleaner and more homogeneous in resin type and scrap form than post-consumer plastics. The term "mixed plastics" has been used to describe broad scale processing of post-consumer plastic waste, although no formal definition yet exists. In its broadest sense, mixed plastics means a collection of a mixture of plastic resins or a mixture of package/product types which may or may not be the same plastic type or color category, and may not have been fabricated using the same manufacturing techniques.

While it is possible to market recycled mixed plastic waste with limited separation, greater value and broader applications are achieved with homogeneous resins. Although it is possible to mix different types of polymers together, the resulting physical properties are less desirable than those of the original components. General strategies for the separation of mixed plastics (and therefore strategies for increased recycling of plastics), with their respective advantages and disadvantages are shown in Table 1.1. Of the five approaches shown, three require substantial technological advances or governmental intervention (or both): "Separation after compaction or shredding," "Container labeling and automated separation" and "Standardization of resin use for certain product applications." Technological research regarding large scale separation of mixed plastic waste streams is being conducted, but some of it is far from commercial application. The advances in plastic separation technology are discussed in Part II of the book.

Governmental regulations regarding standardization for product applications and sortation would most likely not have widespread acceptance. The remaining two approaches, "Manual separation by consumer or collection agency" and "Collection focused on specific resin or container type," are currently in use. They are limited due to the cost of manual labor and a more narrowly defined plastic type, but have the potential for application to wider ranges of plastics than is currently collected by most recycling programs. The purpose of this report is to identify methods used for plastics collection, plastic collection compositions and generation rates, program costs, processing and end market use of mixed plastics in recycling. Attention is given to curbside collection of recyclables due to its high recovery rate[1] versus other collection methods [Morrow and Merriam, 1989]:

Type of Recycling Program	% Recovered
Curbside commingled	70 - 90 %
Curbside home sorted	60 %
Buy-back centers	10 - 15 %
Voluntary drop-off centers	10 - 30 %

---

[1] Recovery rate is defined as the amount of a recyclable disposed in a recycling collection container compared to the total amount of a recyclable disposed by a household.

Table 1.1 Advantages and Disadvantages of Alternative Strategies to Allow Separation of Resin Types from Mixed Recyclable Plastics [U.S. EPA, 1990b]

Strategy	Advantages	Disadvantages
Separation after compaction or shredding	Convenience to consumers; does not require consumers to separate wastes	Currently not possible to separate into homogeneous resins after shredding
	Minimizes sorting, storage and transportation requirements for collecting agencies	Shredding yields mixed plastics not amenable to processing into products displacing virgin resins
	Allows collection stategies capturing large volume and variety of MSW plastics	
Container labeling and automated separation	Convenience to consumers; does not require consumers separate wastes	Technology not currently in place
	Promises to allow separation into homogeneous streams	May require a centralized storage and separation facility with associated costs
	Minimizes manpower requirements required for sorting	
Manual separation by consumer or collection agency	Simple technology	Potentially prohibitive manpower requirements
	Convenience to consumers if collecting agency performs separation	May require large storage and transportation facilities
	Allows collection strategies capturing large volumes of MSW plastics	Inconvenience to consumers if they are required to perform separation

Table 1.1 Advantages and Disadvantages of Alternative Strategies to Allow Separation of Resin Types from Mixed Recyclable Plastics (Continued)

Strategy	Advantages	Disadvantages
Collection focused on specific resin or container types	Facilitates collection of homogeneous resin streams Allows recycling efforts to focus on high-value, high-volume recyclable products Convenience to consumers who are required to collect only a subset of plastic wastes Relatively low cost to recycling agencies Consistent with collection strategies offering financial incentives to recycle	Inconvenience to consumers if they are required to store and transport recyclables to central collection point Captures only a small portion of potentially recyclable plastics
Standardization of resin use for certain product applications	Facilitates collection of homogeneous resin streams	May imply significant governmental intervention in private markets May be difficult to enlist voluntary industry cooperation May be applicable to only a small percentage of recyclable products

1.1 National Production and Recycling Levels of Plastics

Plastic products are found in abundance in the homes we live in, places where we work, clothes we wear, and the transportation we use to get about. Plastics are used to manufacture nondurable goods[2] such as shoes, pens and garbage bags, durable goods[3] such as refrigerators, automotive parts and computers, and packaging such as food tubs, film wraps and bottles. The 1989 U.S. sales of all plastics (including export sales) totaled 58.2 billion pounds, with nearly all of it (92%) being U.S. domestic demand [Modern Plastics, 1990]. Of this, eight plastic types make up the majority of the annual demand (Table 1.2). Six of the types (HDPE, LDPE, PET, PP, PS and PVC) are thermoplastics, capable of being repeatedly softened by increases in temperature and hardened by decreases in temperature. They are also referred to as commodity resins, meaning they are produced in the largest volumes at the lowest cost, and have common characteristics among producers. Polyurethane can be formed as a thermoplastic or a thermoset, the latter of which is a resin which has undergone a chemical reaction leading to a relatively infusible state (that cannot be reformed). Phenolics are a family of thermoset resins.

The major market destinations of plastics production are shown in Table 1.3, with a detailed breakdown of plastic uses in the packaging industry. Consumption of the six thermoplastics is led by the packaging industry (13,568 million pounds), accounting for 36% of the annual demand of the six thermoplastics (37,814 million pounds).

It has been estimated by the U.S. EPA [1990b] that 29 billion pounds of plastic are disposed in the MSW stream each year and that only 1.1% of the waste plastic stream, or 400 million pounds annually, are recovered[4] (Table 1.4). Municipal solid wastes come from residential, commercial, institutional and industrial sources, but do not include wastes such as construction debris, household hazardous waste, or other wastes regulated by Resource Conservation and Recovery Act Subtitle D. Seventy percent of discarded plastic is composed of nondurable goods and packaging materials.

Similar to the EPA estimate of the annual recovery of plastics, a study for the Plastics Recycling Foundation estimated that 340 million pounds were recycled in 1989[5].

---

[2] Nondurable goods are usable only for a short period of time, a lifetime generally less than three years.
[3] Durable goods remain usable for a long period of time, generally products having a lifetime of more than three years.
[4] "Recovery" refers to the removal of materials from the waste stream for the purposes of recycling or composting. This includes materials which may have been removed for recycling, but were stored, landfilled or incinerated due to a depressed market condition.
[5] The primary components of recycled plastics were as follows: 160 million lbs. PET soda bottles, 100 million lbs. HDPE, including 40 million lbs. of soda bottle base cups, 60 million lbs. PP car battery cases, 12 million lbs. PET X-ray film and 10 million lbs. all else [Schut, 1990].

Table 1.2 1989 U.S. Consumption of Leading Plastic Resins

Abbreviation	Resin Type	SPI [a] Code Number	1989 Demand [b] (million pounds)	% of Total [c]
LDPE	Low Density polyethylene [d]	4	9,696	18.1
PVC	Polyvinyl chloride [e]	3	7,564	14.1
HDPE	High density polyethylene	2	7,405	13.8
PP	Polypropylene [e]	5	6,207	11.6
PS	Polystyrene	6	5,037	9.4
PUR	Polyurethane	none	3,245	6.1
none	Phenolic	none	3,140	5.9
PETE or PET	Polyethylene terephthalate	1	1,905	3.6
		Total	44,199	82.6%

a. Society of the Plastics Industry
b. From Modern Plastics, 1990.
c. Percent of total production of all plastics.
d. Includes LLDPE (linear low density polyethylene - 3,286 million lbs) and EVA (ethylene-vinyl acetate - 949 million lbs).
e. Includes copolymers.

This value accounts for only 2.5% of the 1989 domestic packaging demand of the six thermoplastics and less than 1% of all virgin thermoplastic resins. Only one specific bottle type, PET beverage bottles, which have been targeted for recycling through curbside collection and container deposit legislation, has reached notable recycle rates of 23% in 1988 and 28% (175 million pounds) in 1989 [Plastic News, 1990a].

Plastic resins are often difficult to distinguish from one another. Communities performing recycling of plastic containers often train participants to identify a particular container type (such as milk jugs) rather than the actual resin. As an initial answer to this problem, the Society of the Plastics Industry (SPI) established a resin identification system for the six thermoplastics for use in the packaging and container industry (see Table 1.2). The state of Illinois passed legislation requiring the SPI resin identification code on all plastic bottles with a capacity of 16 fluid ounces or more and all other rigid plastic containers with a capacity of 8 fluid ounces or more that are manufactured for use in the

Table 1.3 Major Resin Markets, with Emphasis on Packaging, 1989 [Modern Plastics, 1990]

Market Category			Quantity (10^6 lbs)
Appliances			1,197
Building			11,390
Electrical/Electronics			2,202
Furniture			1,190
Housewares			1,362
Packaging			
	Closures		
		HDPE	81
		LDPE	32
		PP	420
		PS	190
		PVC	36
		Other	20
		sub-total	779
	Coatings		
		HDPE	51
		LDPE	730
		PET	10
		PP	30
		PVC	21
		Other	251
		sub-total	1,093
	Containers		
		HDPE	3,400
		LDPE	311
		PET	1,049
		PP	454
		PS	1,306
		PVC	352
		Other	146
		sub-total	7,018
	Film		
		HDPE	541
		LDPE	3,421
		PP	612
		PS	211
		PVC	310
		Other	107
		sub-total	5,202
Packaging (total)			14,092
Toys			729
Transportation			2,200
Major Market Total			34,362

Table 1.4 Generation, Recovery and Disposal of Plastic Products, 1988 [U.S. EPA, 1990b]

Product Category	Generation (10^6 tons)	Recovery (10^6 tons) (%)		MSW Discards (10^6 tons)
Durable Goods	4.1	< 0.1	1.5	4.1
Nondurable Goods				
Plastic plates and cups	0.4	-	-	0.4
Clothing and footwear	0.2	-	-	0.2
Disposable diapers	0.3	-	-	0.3
Other misc. nondurables	3.8	-	-	3.8
Subtotal	4.6	-	-	4.6
Containers and Packaging				
Soft drink bottles	0.4	0.1	21.0	0.3
Milk bottles	0.4	-	< 0.1	0.4
Other containers	1.7	-	-	1.7
Bags and sacks	0.8	-	-	0.8
Wraps	1.1	-	-	1.1
Other plastic packaging	1.2	-	-	1.2
Subtotal	5.6	0.1	1.6	5.5
Total Plastics	14.4	0.2	1.1	14.3

state beginning January 1, 1991. The code is as follows: PET or PETE - 1, HDPE - 2, PVC - 3, LDPE - 4, PP - 5, PS - 6, and all others - 7. Twenty-six other states, including Indiana, Iowa, Michigan, Minnesota, Missouri, Ohio and Wisconsin have also passed legislation that requires coding by resin type for various plastic containers [Ackerman, 1990]. The increased cost of landfilling waste, the volume occupied by disposed plastic products, the value of the plastic waste material, as well as the mandate of 25% recycling of solid waste by weight set by the state of Illinois make the addition of plastics to recycling programs a necessity. Increasing the recycle of plastic containers, film, and packaging in general, from the waste stream is a logical next step in increasing recycle rates.

1.2 Plastics in Municipal Solid Waste

Recycling of the plastics shown in Table 1.2 has the potential for reducing the waste stream and extending the life of landfills. Currently plastics make up an estimated 9% by weight and 20% by volume of landfill discards [U.S. EPA, 1990b], and most of it (83%) is the six thermoplastics.

The broad identification of disposal routes and types of plastic disposed in MSW landfills on a national basis has been performed by Franklin Associates in a study for the Council for Solid Waste Solutions (CSWS), a program of SPI, which is supported by major petrochemical and polymer production companies. The study examined the disposal routes of the 15 largest resins produced according to 1988 sales and identified which were disposed in MSW and which were not disposed in MSW (Table 1.5). Non MSW-disposed wastes included industrial waste, construction and demolition debris, sludge and incinerator residues. There is a lack of documented information regarding disposal routes of specific plastics and therefore a substantial portion of the research was based on communication with industry manufacturers and resin producers. The data show that for the most part disposal of specific resins is via either MSW or non-MSW methods of disposal (rather than both) and that PVC is the only resin of the leading six that is not disposed predominantly through MSW. Overall, the analysis shows that 61% of plastics are disposed in the MSW stream and 39% in the non-MSW stream. Residences were identified as the primary source of plastics in the MSW stream, comprising 60% of the plastics disposed, followed by the commercial sector contributing 25% and the institutional sector contributing 15% (Table 1.6). The determination as to what plastic products could be apportioned to the three categories of residential, commercial and institutional waste was based on market sales information, grouping of product types, and assumptions on the part of the project team as to where end use of the plastic product would likely occur [Franklin Associates, 1990].

Table 1.5 U.S. Plastic Disposal, 1988 [Franklin Associates, 1990]

Resin	Total Disposed (10^6 lbs)	Non-MSW Disposal (10^6 lbs)	%	MSW Disposal (10^6 lbs)	%
Plastics Disposed Primarily through MSW Route					
ABS	1,093.3	383.1	35.0	710.2	65.0
HDPE	6,528.8	975.3	14.9	5,553.5	85.1
LDPE	7,690.8	577.2	7.5	7,113.6	92.5
PET & PBT	1,475.5	176.2	11.9	1,299.3	88.1
PP	5,274.0	1,016.9	19.3	4,257.1	80.7
PS	4,767.9	529.7	11.1	4,238.2	88.9
Subtotal	26,830.3	3,658.4	13.6	23,171.9	86.4
Plastics Disposed Primarily through non-MSW Routes					
Acrylic	686.0	663.3	96.7	22.7	3.3
Nylon	461.6	329.2	71.3	132.4	28.7
Phenolic	2,975.1	2,869.0	96.4	106.1	3.6
PUR	2,794.8	1,510.7	54.1	1,284.1	45.9
PVC	7,566.0	5,799.4	76.7	1,766.6	23.3
Unsat. Polyester	1,319.3	1,183.0	89.7	136.3	10.3
Urea & melamine	1,459.2	1,346.7	92.3	112.5	7.7
Subtotal	17,262.0	13,701.3	79.4	3,560.7	20.6
Total	44,092.3	17,359.7	39.4	26,732.6	60.6

Table 1.6 Sources of Plastic Disposed in MSW in the U.S., 1988 [Franklin Associates, 1990]

Resin	Quantity Disposed (10^6 lbs)	Residential (10^6 lbs)	%	Commercial (10^6 lbs)	%	Institutional (10^6 lbs)	%
LDPE	7,113.6	4,606.5	64.8	1,641.8	23.1	865.3	12.2
HDPE	5,553.5	3,783.3	68.1	1,007.1	18.1	763.1	13.7
PP	4,257.1	2,138.6	50.2	1,439.9	33.8	678.6	15.9
PS	4,238.2	2,308.0	54.5	1,202.5	28.4	727.5	17.2
PVC	1,766.6	1,033.2	58.5	347.6	19.7	385.7	21.8
PET & PBT	1,299.3	664.9	51.2	286.0	22.0	348.3	26.8
PUR	1,284.1	877.1	68.3	260.9	20.3	146.1	11.4
ABS	710.2	229.0	32.2	370.0	52.1	111.1	15.6
Unsat. Polyester	136.3	68.2	50.0	40.9	30.0	27.3	20.0
Nylon	132.4	98.4	74.3	23.6	17.8	10.5	7.9
Urea & melamine	112.5	88.3	78.5	12.5	11.1	11.7	10.4
Phenolic	106.1	63.7	60.0	32.3	30.4	10.1	9.5
Acrylic	22.7	9.1	40.1	5.7	25.1	8.0	35.2
Total	26,732.6	15,968.3	59.7	6,670.8	25.0	4,093.3	15.3

1.3 Mixed Plastics in Post-Consumer Recycling

Plastics recycling programs usually start with the simplest container/package recognized by consumers and then move on to include additional types. In like manner, classifying post-consumer plastic waste collection could initially begin with a few types and expand to include additional plastic products later. In each recycling program, the consumer/homeowner is taught differently how/what to recycle. This can lead to confusion and inaccurate comparisons between separate recycling programs. It is important to recognize exactly what plastic is being collected if comparisons are going to be made.

The following is a list of plastic types collected in differing curbside recycling programs:

Container / Package Type Collected	Resins Involved
Milk jugs (often includes water & juice)	HDPE clear (unpigmented)
Soda bottles	PET clear or green; HDPE base cup (colored)
All #1 PET items	Any container labeled with SPI 1 (An extension of soda bottle collection)
Beverage bottles	PET, HDPE (Typically first 2 categories)
Detergent and bleach bottles (often includes juice & windshield)	HDPE colored
All #2 HDPE items	Any container labeled with SPI 2 (An extension of milk jug and detergent bottle collection)
All plastic bottles	PET (colored and clear), HDPE (clear and natural), PVC, PP, and some multilayer
All rigid plastic containers (RPC)	PET, HDPE, PS, PVC, PP, multilayer
Any #1 - #7 item	Any plastic labeled with SPI 1 through 7
All RPCs plus film (including plastic bags, film wraps and packaging)	PET, HDPE, LDPE, PS, PVC, PP, multilayer
All clean plastic products	Any plastic emptied of its original contents and rinsed

While there is overlap among the above categories, the list increasingly includes more plastic/product types from top to bottom. Near the bottom of the list, the requirement to identify specific plastic by product type becomes less necessary.

Some recycling programs which are in the middle of the above list (e.g. detergent and bleach bottle category, all plastic bottle category) have lengthy identification/instruction sheets for the homeowner in order to preclude collection of specific containers which are not blow molded (discussed in Part II) or which contain difficult to clean products such as oil containers. The lengthy instruction requirements can lead to non-participation because of homeowner effort and confusion. Confusion can also lead to participants depositing all plastics "just to play it safe." A less confusing approach would be to collect all #1 and #2 bottles, all plastic bottles, or all RPCs, since it has been shown that even a narrowly defined plastic stream (such as plastic beverage bottles) results in a significant portion (>10%) of the plastic deposited not being what was asked for.

2. Characterization, Generation and Collection of Plastics

When considering plastics for inclusion in a recycling program, the questions of "How much plastic is in MSW ?" and "How much plastic has been captured in existing recycling programs ?" come to mind. These questions are even more prevalent when mixed plastics collection is being contemplated. As may be expected, there are few recycling programs which currently collect more than HDPE bottles and PET bottles, and there are even fewer that collect film plastic in addition to rigid plastic containers. While generation rates have been developed for milk jugs and soda bottles based on collection data, there is little collection data for additional plastics. This chapter attempts to provide characterization and generation data from studies and programs involving multiple plastics. Many of the municipalities discussed in this chapter were either participants in pilot mixed plastics recycling studies or received public funding to conduct research on an aspect of curbside recycling collection. The data shown can be used in initial planning for estimating quantities involved in plastics recycling collection and for estimating non-specified plastic and non-plastic contaminants. They can also be compared to the experiences of other municipalities which have conducted forms of mixed plastics recycling. It should be noted that while plastic generation and composition data are presented from national and local studies, it does not replace the need for such an effort in a community's solid waste assessment.

2.1 *Field Assessment of Plastic Types in Municipal Solid Waste*

A study of the composition of plastics in the waste stream in Hamilton, Ohio identified that 6.2% by weight of MSW was plastic. The composition of plastic components is shown in Table 2.1. By comparison, the national average of plastic in MSW discards is 9.2% by weight [U.S. EPA, 1990b]. The "Other" category in the table included such items as HDPE bleach, laundry detergent, motor oil and margarine/butter containers, LDPE band-aids and shampoo bottles, PS caps, and PVC clear bottles [Peritz, 1990]. The Hamilton, Ohio analysis collected 1,000 to 1,200 pounds MSW per day over a 2 week period (11,900 pounds total). Bag/film plastic was the second highest percentage by weight after the "Other" category.

Table 2.1 Plastic in MSW of Hamilton, Ohio [Peritz, 1990]

Plastic Type	% by Weight in MSW	% by Weight of Plastics
PS Foam	0.5	8
PET Soda Bottles	1.0	16
HDPE Milk/Water Bottles	0.7	11
Bags (LDPE film)	1.8	29
Other [a]	2.2	36
Total	6.2	100

a. See text.

The city of Milwaukee conducted a waste characterization during 1989 and 1990 which included the composition of plastics disposed in MSW (Table 2.2). The MSW was from neighborhoods without curbside recycling services. The characterization was based on sorting one load/month of MSW during September, February, May and August to account for seasonal variations. The leaf contribution is zero due to separate collection, and therefore plastics composition should be adjusted. Estimating the contribution of leaves to be 50% by weight of grass clippings results in plastics comprising 7.8% by weight of the total waste stream (rather than 8.4% by weight without including leaves). Film and rigid plastic containers make up the majority of the plastics in the waste stream by weight.

The city of Chicago also conducted a four season waste composition analysis in 1989 and 1990, and identified that 9.4% by weight was plastic, of which 0.8% was PET beverage containers, 1.0% was HDPE containers and 7.6% was other types of plastics [City of Chicago, 1990]. "Other" types included such plastic items as film plastics, durable goods, toys, disposable diapers, packaging and fast-food containers. The data were taken from MSW collected in each of Chicago's 50 wards during April, August, October and February. The average monthly sample size was about 300 pounds per ward.

When assessing waste disposal rates from a large scale municipal facility such as an airport it may be desirable to assess recyclability of the waste stream. A study of plastic composition at Miami, Florida International Airport identified that of the 35 tons/day generated, 12.5 % was plastic by weight. The plastic composition is shown in Table 2.3. Overall, a large percentage of the airport waste was recyclable, with paper, corrugated

Table 2.2 Residential Waste Characterization for Milwaukee with Emphasis on Plastics [Engelbart, 1990]

MSW Component	Composition of Waste Stream (Weight %)
Dirt, diapers and fines [a]	6.6
Ferrous metals	5.1
Food	16.0
Hazardous materials	0.3
Glass	8.3
Leather	0.1
Multi-material packages	0.4
Non-ferrous metals	1.6
Paper	31.0
Plastics	
PET containers $\geq$ 1 liter	0.4
PET containers < 1 liter	0.1
HDPE colored, tubs	0.5
HDPE milk, water, juice, bottles	0.5
Rigid plastics	1.9
Flexible bags and films	4.5
Polystyrene	0.4
Other (durable goods, non-pkg.)	0.1
Subtotal	8.4
Rubber	0.7
Rubble	0.8
Textiles	4.6
Yard waste	
Leaves	0.0
Grass clippings	13.8
Brush / branches / weeds	1.5
Wood (lumber, stumps, pallets)	0.8
Total	100

a. Non-identifiable fine fragments or particles in the waste analysis (often crushed glass).

Table 2.3 Composition of Miami, Florida Airport Plastic Waste Stream [Peritz, 1990]

Product	Weight (%)
Film Plastic	40.6
Clear Cups	16.5
Foams	10.6
Translucent Cups	6.0
PET Beverage Bottles	4.6
Rigid (White) PS	4.4
Utensils	4.4
Rigid Coffee Cups	2.1
Mixed (Other) Plastic	3.3
High Impact Cups (Dairy)	0.8
Straws	0.6
Non-plastic residue	6.1
Total	100

products and glass composing 45.4% of the waste in addition to the plastic. Although this study does not replace the need for such a characterization at Illinois' large municipal facilities, it does provide an indication of what may be expected and demonstrates that the potential exists for large scale recycling at such installations.

A national examination of plastic product types in MSW showed durable goods comprised 29%, nondurable goods 32% and containers and packaging 39% by weight with nearly all of it discarded (Table 1.5, U.S. EPA, 1990b). A materials flow methodology for the waste characterization was used in this study. This method incorporates production data (by weight) for the materials and products in the waste stream with adjustments for imports, exports and product lifetimes. Such a method does not directly allow for apportioning individual waste stream components (such as corrugated cardboard) to differing waste generators (commercial, residential, industrial). Materials flow methodology does not account for product residues present in containers (therefore an estimate of plastic detergent bottles in waste would not include the detergent residue as part of the bottle weight estimate). The method *does* account for the disposal condition of organic wastes such as food waste, yard waste and the moisture present in disposable diapers. The materials flow method is different from site-specific characterization of MSW where a locality may conduct sampling, sorting and weighing of waste stream components. A comparison of the plastic content in MSW using these two independent methods showed plastics were 9.1% (by weight) using the materials flow method and 4.9% - 12.6% (by weight) for 25 site-specific characterization studies [U.S. EPA, 1990b]. Both methods give reasonable estimates.

2.2 Mixed Plastics in Recycling Programs

While percentage of plastics in the waste stream has been studied in a number of locations, less information is available on the type of plastics collected in recycling programs requesting mixed plastics. Each recycling program is unique in its characterization. Most plastic curbside collection is commingled with other recyclables and sorted at a MRF. Curbside collection across the country has only recently started to include plastic containers, and even then most collections are PET and HDPE beverage bottles.

New Jersey

A study by the Center for Plastic Recycling Research (CPRR) at Rutgers University analyzed the curbside and drop-off collection of mixed plastics in which the only plastic types specified were <u>beverage</u> bottles. It showed the bottle composition to be roughly 50% PET soda bottles, 30% milk and water bottles and 10-20% non-specified plastic, by weight [Morrow and Merriam, 1989]. The plastic composition from two of the communities (curbside collection in Mt. Olive, NJ and drop-off collection in South Plainfield, NJ) is shown in Table 2.4. It provides details on plastics contributed that were not specified for collection in the program. There is a large increase in non-specified plastic collection for the drop-off site study in comparison to the curbside collection study. This illustrates the need for public education and clear, simple collection instructions.

In two other municipalities studied by CPRR, the directions for curbside plastic collection were potentially confusing. The handout sheets to residents requested beverage containers only, while instructions printed on the 20 gallon recycling bin requested "all plastic bottles." As a result, non-specified plastics for the two towns averaged 20-25% by weight, about twice the 10.9% of non-specified plastics collected in the Mt. Olive study [Morrow and Merriam, 1990].

Collection of rigid plastic containers (RPC) shows that there is a large constituent beside milk and soda bottles which will be deposited in a recycling bin. A study of the Sayerville, NJ curbside collection program, where plastics accepted included "any plastic bottle or container from which a product is poured," shows HDPE that would be collected in addition to only milk and water bottles (Table 2.5).

In addition to the above municipalities, CPRR has estimated an average generation rate of all recyclables in suburban New Jersey municipalities with single family homes. The average volume and weights of plastic as well as other recyclables expected per household is shown in Table 2.6. The average weekly setout of PET and HDPE bottles (colored and clear) was determined the to be 0.45 lb/setout and 0.30 lb/setout, respectively. The volume data shown can be used as a guide for apportioning truck volumes necessary for curbside sortation. Commingled collection of glass, steel, aluminum and plastics in one container and collection of old newspaper in another container (with processing at a MRF) is recommended by CPRR [Rankin, 1989].

Table 2.4 Composition of Collected Plastic Beverage Bottles [Morrow and Merriam, 1989]

Bottle Type	South Plainfield, NJ (Weight %)	Mount Olive, NJ (Weight %)
Population in collection study	20,000	22,000
Program type	drop-off	curbside
Number of samples	10	11
Time period of collection	4/88 - 1/89	5/88 - 2/89
PET beverage bottles	51.37	55.87
HDPE milk/water bottles	30.10	33.23
Other non-beverage bottles	18.53	10.90
Composition of "Other non-beverage bottles:" [a]		
HDPE detergent bottles	14.74	7.09
HDPE Motor oil bottles	0.63	1.01
PVC bottles	0.75	1.05
PET 2 liter	0.49	0
Other plastic	1.34	1.56
Unknown plastic	0.58	0.19

a. Plastics contributed that were not specified as part of the collection program.

Table 2.5 Composition of Rigid Plastics in Sayerville, NJ Curbside Collection Study [Morrow and Merriam, 1989]

Plastic Collected	Composition (Weight %)	
	3/10/89	4/27/89
PET bottles	41.5	40
HDPE milk and water bottles	15.5	16
Other	43.0	44
Composition of "Other:"		
HDPE Large detergent	22	18
HDPE Small detergent	8	7
HDPE Motor Oil	5	2
Shampoo	3	2
PVC vegetable oil, water	2	5
Miscellaneous	3	10
Total	100 %	100 %

Table 2.6 Composition of Recyclables Collected Weekly per Household in New Jersey [Rankin et al, 1988]

Recycle Stream	Weight (lbs/setout)	Weight (%)	Density (lbs/yd^3)	Volume (gallons/setout)	Volume (%)
Newspaper	7.75	48.4	500	3.1	23.7
Glass bottles	6.0	37.5	700	1.7	13.0
Metal cans	1.0	6.3	144	1.4	10.7
Aluminum cans	0.5	3.1	49	2.1	16.0
Plastic bottles, uncrushed					
PET (60%)	0.45	2.8	40	2.3	17.6
HDPE (40%)	0.30	1.9	24	2.5	19.1
Total per setout	16.0	100%		13.1	100%

Total MSW generated [a] (lb/household/week) = 63.5
MSW recycled (%) = 16.0 / 63.5 = 25.2

a. This is based on 1000 lbs of MSW generated/person/year and 3.3 people per household.

Hennepin County, Minnesota

Studies were conducted in the Minneapolis/St. Paul, Minnesota area in 1990 by the Council for Solid Waste Solutions and the local governments [CSWS, 1990]. The purpose of the work was to assess the feasibility of collecting plastics at curbside in Hennepin County, Minnesota. This study illustrates the addition of various plastics to existing recycling programs using differing collection vehicles and presented details related to composition and collection.

The programs consisted of adding curbside collection of plastics to five existing routes: 3 routes in the city of Minneapolis, 1 route in the city of Minnetonka and 1 route in three northwestern suburbs known as the Hennepin County Recycling Group (HRG). The three Minneapolis routes collected biweekly and the other two routes collected weekly. All methods utilized curbside sortation of material. The study duration was roughly three months. A summary of collection methods for each route is shown in Table 2.7. All field data were collected over a one month period by Cal Recovery Systems, Inc.

Municipal and private haulers were utilized to collect a variety of plastic types. The Minneapolis "A", "B", and "C" routes were operated by the local public works department with differing trucks and plastic collection methods. The Minneapolis "A" route collected plastic soft drink and milk bottles, the "B" route collected all plastic bottles and the "C" route collected all rigid plastic containers. The Minnetonka route collected plastic milk, water, soft drink, and detergent bottles (no bleach) and was operated by Waste Management, Inc. They originally used a cage for plastic mounted in the rear of their Lodal trucks and later changed to nylon bags on the back of the truck. When the bags were full, they were transferred to a standard packer truck dedicated to plastics collection for delivery to the handler. The HRG route collected plastic soft drink and milk bottles and was operated by Browning Ferris Industries. A cage on top of the recycling truck was used for plastics collection.

Compositions and estimates of collection amounts of the different plastics collected from the above described curbside programs are shown in Table 2.8. The specified plastics for the collection area are shown at the top of the table, and the estimated generation per household served per year and proportion of each plastic are shown within the table. The compositions shown do not include non-plastic contaminants. These data should be directly applicable to other moderate to large sized cities located in the midwest.

Table 2.7 Collection Methods Used in Minneapolis Area Pilot Collection of Plastics [CSWS, 1990]

City (Route)	Collector/ Frequency	Number Households Served	Collection Vehicles	Additional Containers Provided	Plastic Types Collected	Duration of Study (weeks)
Minneapolis (Route A)	City Crews / bi-weekly	5,195	Isuzu pick-up with trailer; packer for old newspaper	none	Soft drink & milk bottles	18
Minneapolis (Route B.1)	City Crews / bi-weekly	1,334 [a]	Labrie truck for everything	24 gallon	All plastic bottles	14
Minneapolis (Route B.2)	City Crews / bi-weekly	1,303 [a]	Isuzu pick-up with trailer for everything	24 gallon	All plastic bottles	14
Minneapolis (Route C.1)	City Crews / bi-weekly	1,280 [b]	Labrie truck for everything	24 gallon	All rigid plastic containers	14
Minneapolis (Route C.2)	City Crews / bi-weekly	1,306 [b]	Isuzu pick-up with trailer for everything	24 gallon	All rigid plastic containers	14
Minnetonka	Waste Management / weekly	13,685	Lodal trucks with nylon bag; unload plastics to packer for transport	none	Soft drink, milk, water & detergent (no bleach)	13
Hennepin Recycling Group [c]	Browning Ferris / weekly	3,715	Eager Beaver truck w/ cage	none	Soft drink & milk bottles	13

a. The Minneapolis "B" route originally had a total of 2,543 homes and was later split roughly in half in order to test a prototype perforation device on-board the recycling truck.
b. The Minneapolis "C" route had a total of 2,376 homes and was later split roughly in half in order to test a prototype perforator on-board the Labrie.
c. The Hennepin Recycling Group was a joint program between the towns of Brooklyn Center, Crystal and New Hope.

Table 2.8 Plastic Composition and Collection Amounts in Minneapolis Area Pilot Programs (pounds per household served per year) [CSWS, 1990]

Types of Plastic	Minneapolis Route A	Minneapolis Route B	Minneapolis Route C	Minnetonka Route	Hennepin Route
Collection Area Plastics Specified	Milk & Soft drink bottles	All plastic bottles	All rigid plastic containers	Milk, water, soft drink & detergent bottles	Milk & soft drink bottles
Plastics Collected:					
PET bottles	2.12 (40.8%)	4.90 (27.7%)	4.43 (19.8%)	2.38 (25.4%)	3.25 (36.8%)
PET non-bottles	0.0	0.24 (1.4%)	0.28 (1.2%)	0.01 (0.1%)	0.0
HDPE clear bottles	3.01 (57.9%)	5.74 (32.5%)	5.46 (24.4%)	5.22 (55.8%)	5.44 (61.5%)
HDPE color bottles	0.0	2.62 (14.8%)	3.59 (16.1%)	1.26 (13.5%)	0.0
HDPE non-bottles	0.0	0.15 (0.8%)	0.17 (0.8%)	0.0	0.0
PVC	0.0	0.49 (2.8%)	0.50 (2.2%)	0.03 (0.4%)	0.01 (0.1%)
Composites	0.0	0.23 (1.3%)	0.22 (1.0%)	0.01 (0.1%)	0.0
Unknown bottles	0.0	0.34 (1.9%)	0.83 (3.7%)	0.04 (0.4%)	0.02 (0.3%)
PP	0.0	0.38 (2.2%)	0.23 (1.0%)	0.02 (0.2%)	0.0
PS	0.0	0.19 (1.1%)	0.58 (2.6%)	0.01 (0.1%)	0.01 (0.1%)
Lids & Caps	0.03 (0.5%)	0.55 (3.1%)	1.04 (4.7%)	0.18 (1.9%)	0.05 (0.6%)
Unknown non-bottles	0.0	0.70 (4.0%)	2.48 (11.1%)	0.04 (0.4%)	0.01 (0.1%)
Plastic films	0.0	0.04 (0.3%)	0.01 (0.0%)	0.0	0.0
Non-Plastics	0.03 (0.5%)	1.09 (6.2%)	2.54 (11.4%)	0.15 (1.6%)	0.05 (0.5%)
Total	5.2 (100%)	17.7 (100%)	22.4 (100%)	9.4 (100%)	8.8 (100%)
Plastics portion of recyclables collected (% wt)	3.8%	5.1%	6.5%	3.7%	3.8%

Walnut Creek, California

The results of a 12 week pilot mixed plastic curbside collection program in Walnut Creek, California are shown in Table 2.9. Plastic types collected were any clean film or rigid plastic container. Also shown in the table is the fraction of non-plastic components or "trash." Four samples of 1,300 pounds each were characterized by hand sortation and physical and chemical tests. The average generation rate was reported to be 0.16 pounds/person/day.

Table 2.9 Curbside Collection of Mixed Plastics in Walnut Creek, California [Peritz, 1990]

Product	Composition of Plastics Collected (Weight %)	Range (Weight %)	Trash Compensated [a] (Weight %)
Film	21.0	14.6 - 27.1	24.2
Film Collection Bags	6.5	4.5 - 7.8	7.5
Foam	1.6	0.7 - 2.2	1.8
HDPE Milk/Water Bottles	16.4	11.2 - 19.1	18.9
PET Beverage	9.9	8.2 - 11.7	11.4
Mixed Other Plastics	31.2	27.1 - 34.7	36.0
Trash	13.4	8.0 - 21.2	-
Total	100	-	100

a. Composition of plastics when excluding the "Trash" portion.

Akron, Ohio

Akron, Ohio has curbside collection of rigid plastic containers (RPC) and PS from 14,000 households (1/4 of the city). The commingled RPC collection includes all clean bottles, wide-mouth tubs, trays, miscellaneous packaging and PS foam. Other materials collected are old newspaper, glass, aluminum and ferrous metal. Akron has established a materials recovery facility (MRF) to process residential recyclables and commercial sources of waste plastic, paper and corrugated fiberboard. Plastics are 5% by weight and 25% by volume of the incoming recyclables to the MRF. Plastics comprise 13% by weight of the residential recycle stream. For this municipality, collection of three plastics (HDPE, PET, PS) make up 91% by weight of rigid plastics collected from residences. The composition by weight of collected residential plastics plastics was [Bond, 1990]:

- 57% HDPE
- 28% PET
- 6% PS
- 2% PVC
- 7% mixed plastics

Ontario, Canada

The curbside collection study of RPCs in Barrhaven, Ontario by TransOntario Plastics identified the types of plastic collected using hand sortation and lab analysis. The types of plastic collected were reported by weight as:

- 75% HDPE/PP
- 13% PS/PVC
- 12% PET

Prior to the start of the 24 week Ontario RPC program, no curbside collection of any plastic had occurred. A separate curbside program collected PET beverage bottles (along with other recyclables) prior to the addition of RPCs for the study. Following RPC introduction, the plastics composition by weight was as follows: 65% HDPE/PP, 26% PET and 9% PS/PVC [TransOntario Plastics, 1989].

Seattle, Washington

A six month pilot collection of mixed plastics in Seattle, Washington, which included film, had the following plastic composition by weight:

- 29% HDPE bottles
- 24% PET bottles
- 47% film and "other" plastics.

No further breakdown of the "other" plastic composition was provided. Additional results of the program, which conducted a survey of participants, are described in Chapter 3.

PS Collection Data

The curbside collection of polystyrene is being examined to assess the economics of its recycling. Fitchburg, Wisconsin, a town with mandatory recycling, has conducted a curbside pickup of PS containers in a nine month pilot program funded by Amoco Chemical Company. The town additionally collects old newspaper, mixed paper, any #1 labeled plastic (PET), any #2 labeled plastic (HDPE), aluminum, glass and steel cans. In the first four months of the program, April through September, 1990, 5,281 pounds of PS were collected in a 3,000 household area. The weight of PS includes the weight of plastic bags used to contain the material. Analysis of a 489 pound PS shipment at the reclaimer's facility yielded the following composition of products collected [Adams, 1990b]:

- Container bag weight (lbs/wt. % of total)   67 / 13
- Packing material (wt. %)   25.8
- Meat/vegetable trays (wt. %)   24.2
- Fast food take-out containers (wt. %)   11.8
- Insulation (wt. %)   6.4
- Plates (wt. %)   6.4
- Egg cartons (wt. %)   5.7
- Rigid PS (wt. %)   2.6
- Dirty (moldy) PS (wt. %)   8.5
- Non-PS (wt. %)   8.5

The dirty PS required washing and drying prior to re-use. A sample of one week's normal trash indicated that most residents were using the service because only a small amount of PS was recovered from MSW. According to the town's recycling coordinator, including PS in the collection program does not affect collection truck times or efficiencies. PS is bagged by the homeowner before placement at the curb, and the collector has a separate hammock for placement of the bags. The bags can also be placed in with other plastics depending on the amount set out. Based on the rate of PS recycling thus far, the generation is estimated to be about 5.3 pounds per household per year in the curbside collection area.

2.3 Per Capita Generation

In order to estimate the benefits and costs associated with including plastics in a curbside recycling program, it is first necessary to estimate the amount of plastic recyclables which can be collected. This also helps determine collection truck compartment size and processing area necessary. In mixed plastics, it is helpful to measure the amount of total plastics collected as well as the individual types. With curbside pickup recycling, the setout rate (the percent of households in the collection program which set out recyclables each time collection occurs), the monthly participation rate (the percent of households which set out recyclables at least once in a 4 week period), and the frequency of participation over a time period help ascertain the level of success and degree of participation in a program. For most curbside pickup recycling programs, collection occurs weekly. Often times, only weekly setout rate is recorded by programs and a multiplier is used to estimate monthly participation rate. Even more rare is the recording of the frequency of participation over a period of time and/or recording of these parameters for individual recyclable components such as plastic, HDPE, PET, or PS. The cost of additional labor involved in recording and processing of such data is usually cited as reasons for not performing such a detailed review, and it is typically only done when funded by a grant. Without recording any of the above data, it is only possible to estimate pounds of recyclables collected per household in the collection area. Estimating the pounds per household in the collection area factors in non-participants and does not indicate the true amount set out by a household, but it does give an overall average, assuming participation stays similar. With the frequency of participation or weekly setout rate for each recyclable component recorded, it is possible to determine material per setout and per participating household. This provides the actual contribution on a per setout basis. The frequency of participation also provides the distribution of how often setouts occur.

An example of this can be seen from the recycling experience of Fitchburg, WI. In early 1990, the town received a grant to evaluate homeowner participation regarding each recyclable over an 8 week period, from March 1 to April 26, 1990. The town's weekly curbside recycling program served 2,823 households and collected 11,360 pounds of HDPE for an average of 1,420 lbs./week. The participation assessment study area included 1,185 households and therefore an estimated 596 lbs.of HDPE/week [(1,185 / 2,823)×1,420] were collected from it. Data recorded are as follows:

Total residences in participation assessment area:		1,185	(1)
HDPE collected during the 8 week period (lbs.):		4,768	(2)
Frequency of participation (# residences):			(3)
	1 in 8 weeks	206	
	2 in 8	193	
	3 in 8	178	
	4 in 8	137	
	5 in 8	84	
	6 in 8	73	
	7 in 8	28	
	8 in 8	12	
Total participating households (phh):		911	(4)
Households that never put HDPE out:		274	(5)
8 week HDPE participation rate	= 911/1,185 = 76.9%		(6)

$$
\begin{aligned}
\text{Setouts in 8 week period} &= 1{\times}206 + 2{\times}193 + 3{\times}178 + 4{\times}137 + 5{\times}84 \\
&\quad + 6{\times}73 + 7{\times}28 + 8{\times}12 \\
&= 2{,}824 \tag{7}
\end{aligned}
$$

$$
\text{Ave. setouts/week} = 2{,}824 / 8 = 353 \tag{8}
$$

$$
\text{Ave. weekly setout rate} = 353 / 1{,}185 = 29.8\,\% \tag{9}
$$

$$
\text{Pounds HDPE per setout} = 4{,}768 \text{ lbs} / 2{,}824 \text{ setouts} = 1.69 \tag{10}
$$

$$
\text{Ave setouts/phh for 8 week period} = 2{,}824 \text{ setouts} / 911 \text{phh} = 3.1 \tag{11}
$$

$$
\text{Ave. setouts/phh/week} = 3.1 \text{ setouts/phh} / 8 \text{ weeks} = 0.39 \tag{12}
$$

$$
\text{Ave. time between setouts/phh} = 8 \text{ weeks} / 3.1 \text{ setouts/phh} = 2.6 \text{ weeks} \tag{13}
$$

$$
\begin{aligned}
\text{Est. HDPE/participant household} &= (1.69 \text{ lbs./setout}) \times (52 \text{ wks/yr}) \\
&\quad \times (0.39 \text{ setouts/phh/wk}) \\
&= 33.8 \text{ lbs/phh/year} \tag{14}
\end{aligned}
$$

Therefore, it is estimated that 33.8 pounds of HDPE per year were setout by each house which participated in HDPE recycling. This estimate does not include non-participants. Another way to determine participating household generation is to record and average the weekly setout rates of each recyclable over a period of time. Although simpler and less costly, the disadvantage of this method is that it does not allow a recycling coordinator to view frequency of participation. Viewing the frequency of participation can allow the identification of non-users of a curbside recycling program and can assist in the modification of collection frequencies.

An indication of the plastic generation and recycling rates which can be achieved for the six primary types of plastic are shown in Table 2.10. It shows estimates of generation rates for different types of plastics collection developed from curbside pickup programs and from plastic recycling projection studies. The source of the data is listed along with the calculated per household rates.

For comparison purposes within Table 2.10 of per capita recycling contribution to actual per capita plastic production, the first three line items in Table 2.10 are per capita estimates of plastic consumption. The first item is the national per capita average of all plastics in MSW on a national basis, the second item is the total per capita production of the six primary thermoplastics on a national basis, and the third item is an estimate of the per capita consumption of durable and nondurable plastic products for Massachusetts.

Table 2.10 Estimates of Plastics Generation and Supporting Background Data

Plastic Type	Generation	Collection Size	Set-Out Rate [a] (%)	Participation Rate [b] (%)	Pounds Collected	Source	Comments
All plastics in MSW	117 lb/cap-yr	materials flow methods	-	-	-	EPA, 1990b	National average of all plastics in MSW
Six thermo-plastics	163 lb/cap-yr	material flow methods	-	-	-	CNT, 1990	National average of HDPE, LDPE, PET, PP, PS, PVC production
All plastics	190 lb/cap-yr	material flow methods	-	-	-	Brewer, 1988	Average durable and non-durable plastic product consumption for 1985
All plastic recyclables	24.7 lb/cap-yr	projected	-	-	-	Eyring, 1990	Estimated plastics available for residential curbside-pickup (CSPU) in Chicago area
All plastic recyclables	32-35 lb/hh-yr	projected	-	-	-	Moore, 1990	Estimated CSPU based on 5 sources. PET, PVC, PS, HDPE, LDPE and multi-layer included.
Any plastic (inc. film)	58.4 lb/cap-yr		-	50%	-	Peritz, 1990	Results of a 12 week CSPU mixed plastic study in Walnut Creek, CA
All rigid plastic containers (RPC)	7 lb/cahh-yr [c] 17.2 lb/phh-yr [e]	5,800	23% [d]	51% [d] 92% (all)	17,129	TransOntario Plastics, 1989	Results of a 22 week pilot CSPU of RPCs in Ontario.
All RPCs	20-30 lb/hh-yr	projected	-	-	-	Moore, 1990	Estimated CSPU based on 5 sources. PET, PVC, PP,

Plastic Type	Generation	Collection Size	Set-Out Rate [a] (%)	Participation Rate [b] (%)	Pounds Collected	Source	Comments
All RPCs	22.4 lb/cahh-yr	2,376	61.9%	-	6,520	CSWS, 1990	Results of 14 week bi-wkly CSPU study in Minneapolis. Data collection duration 6wks.
All plastic bottles	16-25 lb/hh-yr	projected	-	-	-	Moore, 1990	Estimated CSPU based on 5 sources. PET, PVC, PP, HDPE (colored and natural) and some multilayer incl.
All plastic bottles	17.7 lb/cahh-yr	2,543	62.3%	-	5,580	CSWS, 1990	Results of 14 week bi-wkly CSPU study in Minneapolis. Data collection duration 8wks.
HDPE all types	9.0 lb/cap-yr	projected	-	-	-	Eyring, 1990	Easily separable HDPE for residential CSPU in Chicago area
HDPE milk & detergent	4.3 lb/cap-yr	projected	-	-	-	Eyring, 1990	Potential for all HDPE residential CSPU in Chicago area
HDPE #2 bottles	16.5 lb/cahh-yr 21.4 lb/phh-yr [f]	2,823	-	-	46,500	Adams, 1990a	1989 results of mandatory Fitchburg, WI CSPU program
HDPE #2 bottles	25.5 lb/cahh-yr 32.9 lb/phh-yr [f]	2,823	29.8% [f] 51.8% (all)	64.2% [f] 82.7% (all)	54,000	Adams, 1990c	Jan.-Sept. 1990 results of Fitchburg, WI CSPU program
HDPE any #2	8 lb/cahh-yr	36,000	42% (all)	83% (all) [g]	144,560	Englebart, 1990	Results of first 6 mo. 1990 CSPU in Milwaukee, WI.

Table 2.10 Estimates of Plastics Generation and Supporting Background Data (Continued)

Plastic Type	Generation	Collection Size	Set-Out Rate [a] (%)	Participation Rate [b] (%)	Pounds Collected	Source	Comments
HDPE bottles PET bottles	24.3 lb/cahh-yr	25,603	-	76% (all)	624,600	TBS, 1989	Results of 6-12 mo. CSPU study of 4 NJ towns. Participation weighted & annualized collection shown.
HDPE milk & detergent (no bleach) PET solft drink	9.2 lb/cahh-yr	13,685	21.5%	-	13,685	CSWS, 1990	Result of 13 wk weekly CSPU study in Minneapolis. Data collection duration 9wks.
HDPE/PET beverage bottles	10-18 lb/hh-yr	projected	-	-	-	Moore, 1990	Estimated CSPU based on 5 sources. Nat. HDPE only.
HDPE milk, PET soft drink	5.2 lb/cahh-yr	5,195	26.1%	-	3,200	CSWS, 1990	Result of 18 wk bi-wkly CSPU study in Minneapolis. Data collection duration 6wks.
HDPE milk, PET soft drink	8.8 lb/cahh-yr	3,715	21.5%	-	8,280	CSWS, 1990	Result of 13 wk weekly CSPU study in Minneapolis. Data collection duration 13wks.
HDPE milk PET soda	10.7 lb/cahh-yr	11,325	24.5% (all)	60% (all) [h]	21,017	Madison, 1990	Results of a 9 week pilot CSPU in Madison, WI
HDPE milk, juice, water	15.6 lb/hh-yr	not reported	not reported	not reported	not reported	Rankin et al., 1988	Average generation of 3 NJ suburbs; 6 mo. duration
HDPE milk bottles	6.5 lb/hh-yr	projected	-	-	-	Fearncombe, 1990	Historical average for Illinois households
HDPE milk bottles	6.6 lb/cahh-yr	36,000	37% (all)	82% (all)	122,902	Milwaukee, 1990	Results of 24 week pilot CSPU in Milwaukee, WI.
HDPE	3-6 lb/hh-yr	projected	-	-	-	Moore, 1990	Estimated CSPU based on

Table 2.10 Estimates of Plastics Generation and Supporting Background Data (Continued)

Plastic Type	Generation	Collection Size	Set-Out Rate [a] (%)	Participation Rate [b] (%)	Pounds Collected	Source	Comments
HDPE laundry bottles	1-2 lb/hh-yr	projected	-	-	-	Moore, 1990	Estimated CSPU based on 5 sources. Colored HDPE
LDPE	6.0 lb/cap-yr	projected	-	-	-	Eyring, 1990	LDPE potential for residential CSPU in Chicago area
PET any #1	4.8 lb/cahh-yr	36,000	42% (all)	83% (all) [g]	87,304	Milwaukee, 1990	Results of 24 week pilot CSPU in Milwaukee, WI.
PET	3.0 lb/cap-yr	projected	-	-	-	Eyring, 1990	PET potential for residential CSPU in Chicago area
PET soda bottles	2.8 lb/cap-yr	projected	-	-	-	Eyring, 1990	Easily separable PET for residential CSPU in Chicago area
PET soda bottles	12.0 lb/hh-yr	projected	-	-	-	Fearncombe, 1990	Average based on household consumption of soda, adjusted for PET soda share
PET soda bottles	6-11 lb/hh-yr	projected	-	-	-	Moore, 1990	Estimated CSPU based on 5 sources
PET #1 bottles	1.9 lb/cahh-yr 4.3 lb/phh-yr [f]	2,823	-	-	5,240	Adams, 1990a	1989 results of mandatory Fitchburg, WI CSPU program
PET #1 bottles	5.0 lb/cahh-yr 11.6 lb/phh-yr [f]	2,823	9.9% [f] 51.8% (all)	28.5% [f] 82.7% (all)	10,600	Adams, 1990c	Jan. - Sept.1990 results of Fitchburg, WI CSPU program

Table 2.10 Estimates of Plastics Generation and Supporting Background Data (Continued)

Plastic Type	Generation	Collection Size	Set-Out Rate [a] (%)	Participation Rate [b] (%)	Pounds Collected	Source	Comments
PET soda bottles	23.4 lb/hh-yr	not reported	not reported	not reported	not reported	Rankin et al., 1988	Average generation of 3 NJ suburbs; 6 mo. duration
PET soda bottles	4 lb/cap-yr	-	-	-	-	Morrow and Merriam, 1990	Total PET soft drink bottles available for recycle
PP	3.5 lb/cap-yr	projected	-	-	-	Eyring, 1990	PP potential for residential CSPU in Chicago area
PS	1.7 lb/cap-yr	projected	-	-	-	Eyring, 1990	Potential for all PS residential CSPU in in Chicago area
PS foam	0.7 lb/cap-yr	projected	-	-	-	Eyring, 1990	Easily separable PS for residential CSPU in Chicago area
PS foam & cont. labeled 6	5.28 lb/cahh-yr	3,000	10.5%	60%	5,281	Adams, 1990b	Results of PS CSPU during 4 month study of participation rates.
PVC	1.5 lb/cap-yr	projected	-	-	-	Eyring, 1990	PVC potential for residential CSPU in Chicago area

a. Set-out rate is specifically for plastic given, unless followed by (all). Weekly collection was used unless stated otherwise, with average taken over time period given in comments.
b. Participation rate is specifically for plastic given, unless followed by (all). Participation rate was monthly, unless stated otherwise.
c. Pounds collected per total number of households in collection area, including non-participants (lb/cahh-yr).
d. Based on a recording of one-truck's route for the last 4 weeks of the program.
e. Pounds collected per participating households (lb/phh-yr), either specifically for plastic participation rate if recorded, or using overall recorded participation rate.
f. Fitchburg, WI mandatory program began 1/1/88. Actual lb/hh-yr set out and participation and set-out rates based on 8 week monitoring between 3/1/90 and 5/1/90 of set-outs contributing particular plastic type.
g. Based on a phone survey of 800 randomly picked households.

Table 2.10 is organized by plastic type collected and shows the recycling parameters when an actual collection study was performed. The setout rates and participation rates shown are for the plastic component specified, unless they are followed by "(all)", in which case the setout rate or participation rate for all recyclables is shown. As can be seen from the table, only a few programs have recorded component specific setout rate or participation rate. The "generation" column indicates *projected* recycling rates on a per capita or per household basis and is based on material flow methods in pounds per household per year (lb/hh-yr), while *actual* recorded quantities from programs show rates on a pounds per collection area household basis (lb/cahh-yr) or a pounds per participating household basis (lb/phh-yr). Generation rate based on participating households provides the best indication of recycling potential. As can be seen in Table 2.10, there is a large range in amount collected, even for similar materials. Urban density, average income level and degree of recycling acceptance in the area are all factors that impact plastic generation rates. However, Table 2.10 does provide an expected range for almost all combinations of plastic types collected.

2.4 *Commercial and Food Sector Sources of Waste Plastic*

In addition to residential post-consumer plastic available for recycling, per capita estimates of post-consumer plastic generated in the food service industry and the commercial sector (aside from food service) have been estimated (Table 2.11). The estimates shown were made using a combination of manufacturing data, measurements from actual recycling programs, and measurements from excavations of landfills or from intercepted MSW [CNT, 1990]. The table shows that HDPE, PS and LDPE comprise a majority of the plastic available for recycling in the U.S. from the commercial and food service sectors.

Examples of additional commercial sources of scrap plastic are shown in Table 2.12. Such materials are usually separated by the source by plastic type. One of the fastest growing areas is the collection of film HDPE and LDPE grocery sacks by grocers. A collection container for the used sacks is typically placed at the entrance to a store for consumers to deposit upon entry. As previously indicated, films form one of the largest components (by weight) of post-consumer plastic waste. This type of collection represents the beginning of broad acceptance of film plastics into recycling programs.

Table 2.11 Projected Plastic in the MSW Stream Available for Recycling in the U.S. [CNT, 1990]

Resin	Total Plastics Produced (pounds/capita-year)	Available for Recycling (pounds/capita-year)			
		Total	Residential	Food Service	Commercial
PS	19.9	5.2	1.7	1.8	1.7
HDPE	33.3	14.0	9.0	4.0	1.0
PET	8.3	3.3	3.0	0.2	0.1
PP	27.7	4.7	3.5	0.6	0.6
PVC	33.2	2.5	1.5	0.2	0.8
LDPE	40.0	14.0	6.0	1.0	7.0
Total	162.5	43.7	24.7	7.8	11.2

Table 2.12 Commercial Sources of Recyclable Plastic Material

Source	Scrap Materials
Airlines, airline food service	PET liquor miniatures, PS cups, PET food service items.
Bakeries	PP bread trays, tote bins
Dairies	Scrap HDPE bottles, packing crates
Drinking water distributors	PC 5 gallon water bottles, PP crates
Dry cleaners	LDPE garment bags
Equipment/parts manufacturers	Tote boxes, shipping crates, scrap
Farming enterprises	PE agricultural film, feed, fertilizer bags
Food service/cafeterias	Foam PS trays, PS utensils
Grocers	LDPE grocery sacks, HDPE grocery sacks
Large construction sites	PE shrink wrap, construction film
Newspaper distributors	LDPE home delivery newspaper covers
Nurseries	HDPE and PP buckets and pots, PS plant trays
Soft drink bottlers	Deposit bottles, PP crates

2.5 Post-Consumer Plastic Weights

Tables 2.13 and 2.14 show typical properties of collected HDPE and PET, and other recyclables. The pellet and flake values shown for milk bottles may be used for other blow molded HDPE bottles (such as detergent) as well. The "flake" values represent material conditions following grinding. Such detailed information does not readily exist for other plastics collected in recycling programs.

2.6 Summary

This section has presented data on plastics generation. Plastics comprise 6 to 9% by weight of MSW. Depending on the material collected, plastics will comprise 4 to 14% by weight of the recyclables in a recycling program. When collected, film plastics make up a large portion (25 to 40% by weight) of plastics collected. Collection of any type plastic bottle will result in a 75 to 200% by weight increase in the amount of plastics collected over collection of just PET and clear HDPE beverage bottles.

There is a large range of amounts of plastics collected in plastics recycling programs. In order to accurately assess the effect of plastics on a recycling program, a municipality specific composition estimate should be conducted. Generally, rigid plastic recycling will capture 30 lb/hh-yr, plastic bottle recycling will capture 20 lb/hh-yr, and beverage bottle recycling will collect 15 lb/hh-yr, as a minimum, on a per household served basis.

The contribution of non-specified plastics and non-plastic contaminants to plastics recycling programs is dependent on the plastic types collected. Non-specified plastics range from as little as 1% for relatively simple plastics collection (PET and clear HDPE beverage bottles) to as much as 10 to 20% for plastic bottle or colored HDPE recycling. Non-plastic contamination can be 1% to 10% by weight. Even a narrowly defined plastic stream of PS resulted in 8% non-PS plastics and non-plastic contaminants.

Table 2.13 Typical Properties of HDPE and PET Beverage Bottles [PRC, 1990a]

Description	2 liter PET beverage bottles	1 gallon HDPE beverage bottles
Weight / bottle (lbs)	0.14	0.15
Bottles / lb. (number)	7.1	6.1
Bulk density, uncrushed (lbs/ft^3)	1.5	0.9
Bulk density, stepped-on (lbs/ft^3)	3.0	1.8
Typical quantities for a gaylord size of 34"x43"x38"		
Uncrushed, weight / gaylord (lbs)	48	29
Stepped-on, weight / gaylord (lbs)	96	58
Uncrushed, bottles / gaylord (number)	341	194
Stepped-on, bottles / gaylord (number)	682	389
Typical quantities for a bale size of 31"x45"x63"		
Weight / bale (lbs)	750	600
Quantity / bale (number)	5,325	4,020
Target minimum for shipping (lbs/ft^3)	15[a]	12[b]
Processed resin properties		
Pellet density (lbs/ft^3)	50	35
Flake density, 5/16" max size (lbs/ft^3)	29	27
Flake density, 3/8" max size (lbs/ft^3)	28	26
Typical flake and pellet weights for a gaylord size of 34"x43"x38"		
Pellet	1,600	1,060
Flake, 5/16" max size (lbs)	930	870
Flake, 3/8" max size (lbs)	900	840

a. 15 lb/ft^3 specified by Plastic Recycling Corporation of California to achieve 40,000 lb semi loads.
b. 12 lb/ft^3 realistic target, with typical range of 8-15 lb/ft^3

Table 2.14 Specific Volumes and Densities for Recyclable Waste [Rankin et al, 1988, Moore, 1990]

Material	Specific Volume (yd^3/ton)	Bulk Density (pounds/yd^3)
Newspaper	3.3	500-600
Baled shredded paper bundles	2.7	740
Whole glass bottles	3.3	600-700
Clear and colored glass, -5/8" cullet	0.9	2200
Clear and colored glass, -2" cullet	2.0	1000
Whole ferrous cans	10-13	145-200
Flattened ferrous cans	2.2-2.5	800-910
Whole aluminum cans	27	49-74
Flattened aluminum cans	8	250
Uncrushed 1 gallon HDPE milk and water containers	93	22-24
Uncrushed HDPE detergent/bleach bottles	67	30
Hand crushed 1 gallon HDPE milk bottles (74 g/bottle)	76	26
Crushed HDPE detergent/bleach bottles	33-40	50-60
Uncrushed 2 liter PET bottles	58	34
Hand crushed PET 2 liter bottles	41	49
Baled PET	4.3	460
Chipped PET	2.5	800
Uncrushed mixed bottles	57	35
Crushed mixed bottles	25	80
Uncrushed rigid plastic containers	44-50	40-45
Crushed rigid plastic containers	22-25	80-90
Uncrushed any type plastic	15	135
Crushed any type plastic	13	160

3. Plastics Recycling Programs

Curbside recycling is growing at a tremendous rate. In 1989, it was estimated that 9 million U.S. households were part of curbside recycling programs and that 20% collected some type of plastic. In 1990, about 40 million people (roughly 14 million households) were participants in curbside recycling [Glenn, 1990]. By 1992, 20% of all U.S. households (16 million households) are expected to have curbside recycling and 180-330 million pounds of plastic will be recycled annually [COPPE].

3.1 Curbside Collection of Plastics in Illinois

As of August 1990, approximately 110 municipalities conducted curbside collection of recyclables in Illinois. It is estimated that 600,000 households will be served by curbside recycling by the end of 1990 [Fearncombe, 1990]. A review of these municipalities indicates 43 collect some type of post-consumer plastic affecting a total of 221,000 residences. This represents a dramatic increase since 1987, when only a few communities collected plastics. Curbside plastic recycling involvement in Illinois can be one of three types: programs which collect a form of mixed plastic (Table 3.1), those which collect only PET and natural HDPE beverage bottles (Table 3.2), and those which collect only natural HDPE bottles (Table 3.3). Approximately 22 municipalities (87,500 residences) collect a form of mixed plastics, 5 municipalities (27,600 residences) collect PET and natural HDPE beverage bottles and 16 municipalities (106,100 residences) collect natural HDPE bottles. The mixed plastics curbside collectors, for the most part, collect blow molded colored and natural HDPE and PET bottles, as shown in Table 3.1. The "Plastics Collected" column in Table 3.1 attempts to report as accurately as possible the plastics requested of residents. The Naperville Area Recycling Center (NARC) is the widest ranging mixed plastic recycling program, additionally collecting PS and LDPE 6-pack rings as part of test programs for Amoco and Illinois Tool Works, respectively.

3.2 Film / Rigid Plastics Recycling

A number of areas, especially on the east coast, are now collecting RPCs. Some are pilot programs in test areas and others are fully implemented. Large plastic processors such as Wellman (Allentown, PA), Waste Management (Oak Brook, IL), Day Products (Bridgeport, NJ) and Union Carbide (Pistacaway, NJ) are planning large scale RPC curbside collection and processing in the next few years. The most locally notable programs are the Chicago Park District Plastics-on-Parks drop-off program which accepts RPCs, and the curbside collection of rigid plastics in Akron, Ohio. A few recycling

Table 3.1 Illinois Recycling Programs Collecting Mixed Plastics

Municipality	Plastics Collected	Number of Curbside Pickup Households	Collector	Start of Plastics Collection	Number of Drop-off Centers
Alsip	All RPCs [a]	3,000	Groen Bros.	5/90	0
Arlington Heights	PRA mix [b]	10,032	Waste Management	9/88 (HDPE) 1/90 (PET)	2
Bloomington	#1 bottles #2 bottles	7,979	Own and Browning Ferris	1/89 (milk) 9/89 (soda) 7/90 (#1, #2 bottles)	
Buffalo Grove	PRA mix	8,400	Waste Management and Own (drop-off)	7/90	2
Darien	HDPE colored and clear bottles PET soda	~8,000	Rot's Disposal	7/90	0
DeKalb	HDPE colored and clear bottles PET soda	~9,000	DeKalb Disposal	12/88	1
Dixmoor	All RPCs	600	Groen Bros.	4/90	0
Elk Grove Village	PRA mix	~9,100	Waste Management	11/88 (milk) 7/90 (other)	0
Elmhurst	#1 bottles [c] #2 bottles	~1,500	Municipal Recycling and Browning Ferris (drop-off)	4/90	1
Freeport	#1 bottles #2 bottles	3,400	Moring Disposal	7/90	1
Fulton	#1 bottles #2 bottles	1,452	Moring Disposal	9/90	0
German Valley	#1 bottles #2 bottles	170	Moring Disposal	8/90	0
Lake in the Hills	#2 bottles	1,900	Valley Sanitation	7/90	0
Lansing	HDPE colored and clear bottles PET soda	~8,000	Own	1/90 7/90	0
Morrison	#1 bottles #2 bottles	1,800	Moring Disposal	11/90	0
Naperville	HDPE bottles PET soda 6-pack rings PS	25,000	Naperville Area Recycling Center (NARC)	1/90 (6-pack rings) 10/90 (PS)	1
Oregon	HDPE and PET bottles	1,200	Moring Disposal	3/90	0
Palos Heights	All RPCs [a]	4,000	Groen Bros.	5/90	0

Table 3.1 Illinois Recycling Programs Collecting Mixed Plastics (Continued)

Municipality	Plastics Collected	Number of Curbside Pickup Households	Collector	Start of Plastics Collection	Number of Drop-off Centers
Palos Park	All RPCs [a]	2,000	Groen Bros.	5/90	0
Princeton	#1 bottles #2 bottles	2,300	Local center	9/88	1
Rolling Meadows	PRA mix	5,300	Waste Managment	7/90	1
Sycamore	HDPE colored and clear bottles PET soda	200	DeKalb Disposal	12/88	1
Sterling	Any #1 or #2 (inc. tubs)		Rock Valley Disposal	7/90	
Wheaton	HDPE colored and clear bottles	12,000	Waste Management and NARC (drop-off)	8/89	1
Wheeling	PRA mix	5,000	Waste Management	7/90	0
Woodridge	HDPE colored and clear bottles	none for plastic	NARC	4/88	1

a. Any RPC which did not previously contain a chemical.
b. Plastic Recycling Alliance (PRA) mix is defined as unpigmented HDPE milk, juice, water and windshield wiper fluid bottles, colored HDPE bleach, laundry detergent and fabric softener bottles, and clear and colored PET $1/2$, 1, 2 and 3 liter soda/beverage bottles.
c. No bleach or motor oil bottles.

Table 3.2 Illinois Curbside Recycling Programs Collecting Only Natural HDPE and PET Beverage Bottles

Municipality	Start of Plastics Collection	Number of Curbside Pickup Households	Collector	Number of Drop-off Centers
Barrington	10/90	3,500	Browning Ferris	1
Geneva	5/90 (HDPE) 10/90 (PET)	4,214	Speedway Disposal	0
Skokie	7/90	17,600	Haulaway	1
Mt. Carmel	10/88	1,300	K/C Disposal	0
Quincy	9/89 (HDPE) 2/90 (PET)	4,500	Own	0

Table 3.3 Illinois Curbside Recycling Programs Collecting Only Natural HDPE Bottles

Municipality	Start of Plastics Collection	Number of Curbside Pickup Households	Collector	Number of Drop-off Centers
Chicago	11/89	12,445	Own	
Collinsville	4/89	8,900	Laidlaw	0
Downer's Grove	5/90	12,550	Browning Ferris	1
Edwardsville	9/89	4,200	Laidlaw	0
Glenn Ellyn	1/88	~9,000	NARC [a]	1
Glen Carbon	9/89	2,600	Laidlaw	0
Golf	-	200	Laidlaw	0
Hanover Park	8/89	8,500	Laidlaw	0
Highland Park	10/89	8,500	North Shore Waste	0
Homewood	8/89	6,125	Homewood Disp.	0
Oakbrook	3/90	2,600	Oakbrook Disp. (84%)[b] Waste Management (8%) Rot's Disposal (8%)	0
St. Charles			Waste Management	
Schaumburg	7/89	11,500	Laidlaw	0
Streamwood	7/89	9,500	Laidlaw	0
West Chicago	7/89	9,500	Laidlaw	0

a. Naperville Area Recycling Center
b. Percentage relates to market share.

programs are collecting any rigid/film plastic. Most of these are located in close proximity to a plastic lumber manufacturer which accepts mixed plastic film or commingled bales. This includes communities around Toronto, Ontario, on Long Island, New York, in central Michigan and in northwestern Iowa.

Milwaukee, Wisconsin started collecting any type of plastic with an SPI 1-7 label in August, 1990 from a 5,147 residence pilot area and has expanded since. The city also collects HDPE milk bottles and other recyclables from 40,000 additional households as part of its established curbside recycling program. Waste plastic contribution from the pilot area has been strong. The flyer sheet distributed to residents is shown in Figure 3.1. The city uses semi-automated side loading trucks with partial commingled collection. Newspaper, cardboard and magazines are emptied into a bin separate from other recyclables. Material is sorted out on a conveyor belt at a MRF with manual picking stations. HDPE, PET and PVC are picked out and baled and the remaining plastics are baled commingled. Per capita generation and the cost of adding mixed plastic collection have not been estimated.

There have been pilot RPC/film collection programs with poor results due to extensive (uneconomical) sorting necessary at a materials recovery facility (MRF) to prepare material for market and due to food contamination. One such program, conducted in the Portland metropolitan area in 1990, had high amounts of food and residue contamination which resulted in the landfilling of large amounts of film and containers. There was also a large reject rate (30%) due to unacceptable materials, e.g. rubber hoses, household medical waste. Consumers reported difficulty in having to believe it was necessary to sort out all plastics at a MRF to obtain any market price. Resident education as a key element, including cleaning and proper preparation of recyclable plastic, was stressed multiple times as a solution to resolving the problem. Although residents were asked to change patterns in update mailings, participants from the beginning of the program did little to change their patterns. The sortation company participating in the pilot indicated that the types of plastic to be collected should have been more limited.

A six month pilot collection of mixed plastics was conducted on 4,500 households between November, 1989 and April, 1989, in Seattle, Washington. The program, which included collection of film and plastic wrappers in addition to more commonly collected plastics, allowed the following plastics to be disposed in curbside collection as described by the mailer to participant homeowners:

Containers:
- Pop and water bottles
- Milk and juice jugs
- Dishwashing soap bottles

Plastic Bags:
- Grocery bags (produce & checkout kind)
- Food bags (bread, popcorn, cereal, etc.)
- Trash can liners

RECYCLING UPDATE

Thanks to your support, Milwaukee's Recycling Program continues to be successful. The State of Wisconsin passed a major recycling law in April, 1990. Soon we will not be able to throw any glass, plastic, paper, steel or metal cans in landfills and we will have to recycle these items.

To give us an idea of how much of these materials we will be collecting, we are starting an expanded collection in your area. As of **December 3, 1990**, you can put the following in your bin for pickup each week.

1) **PLASTIC CONTAINERS**--You can put any plastic container in your recycling bin IF it has the recycling symbol on the bottom with a number of 1-7.

PETE HDPE V LDPE PP PS OTHER

2) **FOAM POLYSTYRENE PRODUCTS** such as fast food clamshells and meat trays. **PLEASE RINSE!**

3) **CARDBOARD**--We will also pick up cardboard. **IT MUST BE CUT UP INTO PIECES NO LARGER THAN 2' X 2' AND STACKED NEATLY UNDER THE RECYCLING BIN FOR COLLECTION OR PUT IN A PAPER BAG.**

4) **MAGAZINES** can be included in the same bundle or bag with cardboard.

So, as of **December 3, 1990**, you can recycle the following items:

TYPE OF RECYCLABLE	HOW TO PREPARE
Glass containers	any colors, rinse out, no lids
Aluminum cans	rinse out
Tin cans	rinse out, throw lids in garbage, leave labels on
Plastic containers	rinse out, check bottom for 1-7, throw lids in regular garbage, step on containers to flatten
Foam polystyrene (clamshells)	rinse out
Newspaper	put in paper garbage bag or tie with twine
Cardboard	cut boxes into 2'x2' pieces, put under bin or in separate bag
Magazines	put with cardboard or in separate bag

♻ printed on recycled paper

WASTE REDUCTION AND RECYCLING PROGRAM
City of Milwaukee • Department of Public Works • 841 N. Broadway • Milwaukee, WI 53202 • 414-278-3500

Figure 3.1 Flyer Sheet Distributed to Mixed Plastic Collection Pilot Program Residents in Milwaukee, WI.

<u>Containers (cont.)</u>:
- Laundry detergent bottles
- Shampoo and lotion bottles
- Food containers

<u>Plastic Bags (cont.)</u>:
- Merchandise bags from stores
- Dry cleaner bags

<u>Wrappings</u>:
- Wrappings from around toys, razors, hair brushes, utensils, etc., without paper backing
- Wrappings from paper towels, toilet paper, cosmetics.

The Seattle program was the most aggressive mixed plastic collection program ever initiated. In addition to the curbside pickup, drop-off boxes were placed at two transfer stations and at two local stores. The drop-off boxes ended up being the most successful and accounted for 76% of the 161,000 lbs. collected over the six month program.

The curbside service, which was divided into two collection areas (north and south), had extremely different responses. The north end (1,000 customers) collected 25,300 lbs. plastic, while the south end (3,500 customers) collected 13,594 lbs. plastic. Possible explanations for this large difference was the collection frequency for the south end (monthly) compared to the north end (weekly) and the different socieconomic structure between the two areas. Another difference is that a greater percentage of people in the south end surveyed (40%) thought that the plastic was "too messy," as compared to the north end (30%).

The subcontractor, which was to ship the material to Thailand for processing, could not provide a viable long term market for the mixed plastics collected, and therefore the film and wrapper portion of the project was not continued beyond the end of the pilot. Since it was determined that sufficient markets existed for plastic HDPE and PET bottles, this part of the pilot project was extended for three months. Seattle currently collects PET soft drink bottles city-wide.

The Seattle program conducted a survey of 78% of the participants to determine willingness to recycle various types of plastic. The return rate was 43%, which resulted in an estimated margin of error of $\pm 1.5\%$. Willingness to recycle plastic types was as follows: plastic bags (87%), food bottles (79%), milk jugs (54%) and soft drink bottles (43%) [City of Seattle, 1989]. Of the participants surveyed, 16% indicated the project made them cut down on plastic, while 11% bought more. Direct mail was an effective way to communicate with the pilot program audience with 86% of the respondents recalling the letter asking each respective residence to participate. Other promotional mail material had high recall rates as well. The survey also determined that preparation of the recyclables did not act as a barrier in plastics recycling.

4. Recycling Costs

It is difficult to address the cost of plastic collection and processing without including other recyclables as well. How effectively plastics can be added to a recycling program depends on the current collection system and the flexibility of it. It is necessary to evaluate these issues in order to identify the best method for meeting recycling goals and for providing an indication of capital and operating budget expenditures. Because there are large variations between levels of recycling, it is almost always necessary to examine recycling program costs on a case-by-case basis.

Presented in this chapter is a cost estimate of curbside recycling options for the City of Madison, Wisconsin and an estimate of the cost of <u>adding</u> differing levels of plastics collection to existing curbside collection programs in the Minneapolis/St. Paul, Minnesota area. Each is important because they examine a number of options, thereby providing a range of values based on the options, and because their estimates include the utilization of existing equipment, the need for capital purchases is minimized. Also presented are costs which may be expected for processing plastics (baling, sorting, grinding) at a material recovery facility, and computer methods for estimating and optimizing recycling costs.

4.1 Recycling Program Variables

Because curbside pickup achieves the highest recovery rates, cost effectiveness of recycling was evaluated using curbside collection. The following variables affect curbside or drop-off collection costs:

- Recyclables collected
- Method of sorting (curbside versus MRF)
- Households in the service area
- Household participation rate
- Collection period setout rate
- Collection frequency
- Portion of recyclables separated by the homeowner for recycling (termed capture rate)
- Generation per person or household
- Travel time between households
- Time required per stop
- Travel time to dump
- Time at the processing center
- Cost of personnel and equipment
- Personnel per truck
- Market prices of recyclables
- Cost to process/level of marketability
- Shipping cost
- Cost avoidance of landfill diversion

The generally accepted method for determining the cost of recycling is:

Recycling cost = Revenue from sale of recyclable material
(or profit)   + Avoided cost of MSW collection
              + Avoided cost of MSW disposal
              - Cost of collection of recyclables
              - Cost of sortation at MRF
              + Future value of saved landfill space

4.2 Recycling Costs

Although there is a social desire to recycle, the current price of landfill space in some areas has not yet offset the cost of curbside collection of recyclables. Recycling in Illinois is generally paid for on a per household basis. The cost of recycling may be expected to add 10-25% to the cost of existing refuse disposal. This translates into an additional cost of $1-$2.50 per month per household, either paid by a municipality throu general/taxpayer funds or directly included on homeowner bills.

Madison, Wisconsin Recycling Cost Estimate

An example cost estimate of recycling options was performed by the city of Madison, Wisconsin in 1990. A pilot study was conducted because the local county landfill is facing closure, and the local legislature banned disposal of certain recyclables ii the landfill after January 1, 1991. The following options were considered for the mandatory recycling cost estimates:

- Option 1   Once/week collection on weekends only, using recycling bags (rather than bins) which would be purchased by residents. Existing refuse collection equipment would be used (rear load/side load packers).

- Option 2   Once/week collection on weekdays only, using recycling bags which would be purchased by residents. Specialized collection equipment would be purchased.

- Option 3   Once/week collection on weekdays only, using recycling bins which would be purchased by the city. Specialized collection equipment would be purchased.

- Option 4   Same as option 1, except bi-weekly collection.

- Option 5   Same as option 2, except bi-weekly collection.

- Option 6   Same as option 3, except bi-weekly collection.

The assumptions used in the cost estimate are shown in Table 4.1. The cost comparison between each option is shown in Table 4.2. As may be expected, the cost for

Table 4.1 Assumptions Used to Calculate Costs of Alternative Curbside Recycling Options for Madison, Wisconsin [City of Madison, 1990]

Option	1	2	3	4	5	6
Equipment Collection device	Existing Bag	Special Bag	Special Bin	Existing Bag	Special Bag	Special Bin
Collection frequency	Weekly	Weekly	Weekly	Bi-Wkly	Bi-Wkly	Bi-Wkly
Collection days	Weekend	Weekday	Weekday	Weekend	Weekday	Weekday

Decision parameters

	1	2	3	4	5	6
Participation rate (%)	70	70	80	70	70	80
Set out rate (%)	31.2	31.2	40.0	44.6	44.6	58
Stops/week	15,600	15,600	20,000	11,166	11,166	14,500
Collection days	Sat&Sun	M-F	M-F	Sat&Sun	M-F	M-F
Pounds/stop	10.3	10.3	8.8	14.3	14.3	12.1
Collection time/stop (sec)	54.2	48	52	50.9	45.6	49.4
Equipment capacity (lbs)	5,000	4,200	4,200	5,000	4,200	4,200
No. stops/crew/day	332	375	347	350	300	347
Crew days/week	47	41.6	58	32	31.5	41.8
No. days operated/week	2	5	5	2	5	5
No. crews/day	23.5	8.3	11.6	16	6.3	8.4
Additional foremen needed	1	0	0	1	1	1
Additional equipt. operators	10	9.5	13	8.5	7	9.5
Additional trucks needed	0	10	13	0	8	10
Cost/truck ($)	110,000	67,500	67,500	110,000	67,500	67,500
Debt service/truck	8,164	9,450	9,450	8,164	9,450	9,450
Route (mi./truck/day)	40	42	40	40	44	45
Equipt. oper. cost ($/mi.)	0.38	0.20	0.20	0.38	0.20	0.20
Equipt. maint. cost ($/mi.)	1.00	0.29	0.29	1.00	0.29	0.29
Add'l bldg. space needed (ft^2)	0	14,500	18,600	0	14,450	14,450
Container cost	$.10/bag	$.10/bag	$7.50/bin	$.10/bag	$.10/bag	$7.50/bin
Container life	One use	One use	5 years	One use	One use	5 years
No. containers used/year	26	26	1.3	20	20	1.3

54 Mixed Plastics Recycling Technology

Table 4.2 Annual Cost Comparison of Curbside Recycling Alternatives for Madison, Wisconsin [City of Madison, 1990]

Option	1	2	3	4	5	6
Equipment	Existing	Special	Special	Exisitng	Special	Special
Collection device	Bag	Bag	Bin	Bag	Bag	Bin
Collection frequency	Weekly	Weekly	Weekly	Bi-Wkly	Bi-Wkly	Bi-Wkly
Collection days	Weekend	Weekday	Weekday	Weekend	Weekday	Weekday
Cost Category						
Labor						
Overtime Salaries	0	0	0	7,543	7,740	10,085
Permanent Salaries	252,028	192,205	269,859	193,557	191,009	242,905
Benefits	75,238	61,331	83,296	61,587	60,952	77,560
Sub-total Labor	$327,266	$258,536	$353,785	$255,144	$251,961	$320,465
Materials, Supplies & Purchase Services						
Safety	1,000	1,000	1,300	700	800	1,000
Tools	0	550	650	0	400	500
Service Vehicle	15,000	0	0	15,000	0	0
Containers-buy & dist.	0	0	100,000	0	0	100,000
Sub-total mat. & purch	$16,000	$1,550	$101,950	$15,700	$1,200	$101,500
Motor Equipment						
Debt Service	195,936	94,500	122,850	130,624	75,600	94,500
Operation	37,149	17,264	24,128	25,293	16,380	17,472
Trasnsportation	85,540	26,910	35,648	66,560	24,079	21,840
Sub-total equipmenrt	$318,625	$138,674	$182,626	$222,477	$116,059	$133,812
Vehicle storage facility						
Debt service	0	111,265	143,220	0	111,265	111,265
Operation & maintenance	0	15,000	20,000	0	15,000	15,000
Sub-total buildings	$0	$126,265	$163,220	$0	$162,265	$126,265
Sub-total Operations	$661,891	$525,025	$801,581	$493,321	$495,485	$682,042
Public education	50,000	50,000	50,000	50,000	50,000	50,000
TOTAL Costs	$711,891	$575,025	$851,581	$543,321	$545,485	$732,042
Less:						
Tip fee diversion	($72,460)	($72,460)	($79,900)	($72,460)	($72,640)	($79,900)
Material revenue	0	0	0	0	0	0
Sub-total revenues	($72,460)	($72,460)	($79,900)	($72,460)	($72,640)	($79,900)
Net Total Costs	$639,431	$502,565	$771,681	$470,861	$472,845	$652,142
Tons Recycled	4,200	4,200	4,565	4,200	4,200	4,565
Net cost/ton	$152.25	$119.66	$169.04	$112.11	$112.58	$142.86
Cost/household/month	1.07	0.84	1.29	0.78	0.79	1.09
HH. cost/container/month	0.22	0.22	0.17	0.17	0.17	0.17
Total Cost/hh/month	$1.29	$1.06	$1.46	$0.95	$0.96	$1.26

once/week collection was higher than bi-weekly collection. Overall, cost estimates ranged from $0.96/month/household to $1.46/month/household. None of the options included the cost of processing or the revenues generated from the sale of recyclables.

In order to assess the cost of sorting and processing the recyclables collected, data from the nine week pilot study can be utilized. During the pilot, two contractors were paid $65.18/ton to process and separate the recyclables, and revenues of $42.20/ton were generated, resulting in a net cost to the city of $22.98/ton for recycle processing. Using this net processing fee and extrapolating the 126.25 tons of recyclables collected for the 11,325 household pilot to a yearly basis would result in an additional processing cost of $1.48/household/year. This example provides an objective cost estimate of implementing a municipal recycling program for a large community (50,000 households) and evaluates the use of new and existing equipment. A reasonable estimate of cost for municipalities greater than 50,000 households may be determined by scaling the values accordingly.

Minneapolis, Minnesota Plastic Recycling Cost Estimate

The Council for Solid Waste Solutions (CSWS) conducted a collection analysis for the Minneapolis, MN area in 1990. The impact of plastics collection on five existing curbside collection recycling programs was evaluated in terms of additional collection time, vehicle capacity and additional cost. The collection routes were as follows:

- Minneapolis A   - Clear HDPE and PET soda bottles
- Minneapolis B   - All plastic bottles
- Minneapolis C   - All rigid plastic containers
- Minnetonka   - Clear HDPE, colored HDPE, PET soda bottles
- Hennepin Recycling Group - Clear HDPE and PET soda bottles

Four different collection vehicles were used:

- An *Eager Beaver* brand recycling truck with a cage mounted on top for holding plastics, used on the Hennepin Recycling Group (HRG) route

- A *Lodal* brand recycling truck with nylon bags mounted for holding plastics, used on the Minnetonka route

- A *Labrie* brand semi-automated side loading recycling truck with an on-board perforator and flattener to densify plastics, used on Minneapolis routes B and C

- A customized *Isuzu* truck and trailer system with divided compartments for recyclables (including plastics), used on Minneapolis routes A, B and C

The following biweekly plastic set-out rates for Minneapolis measured by the recycling contractors were reported: Minneapolis A - 26.1%, Minneapolis B - 62.3%, Minneapolis C - 62.1%. Weekly plastic setout rates for Minnetonka and HRG were 21.5% and 29.5%, respectively. The Minneapolis B and Minneapolis C plastic setout rates were most likely so significantly higher than the setout rates of the other routes because they collected any type plastic bottle and any rigid plastic container, respectively.

A 50,000 household cost estimate of each of the three Minneapolis plastic recycling options was conducted using data from each of the routes. The estimate factored in amortized cost of capital expenses, administration, operation and maintenance expenses, labor expense based on collection times recorded, and waste diversion credit for the Minneapolis area. A summary of the economic analysis using data collected in the three Minneapolis (A, B and C) programs is shown in Table 4.3. For Minneapolis, adding plastics to a biweekly recycling program using semi-automated side loading trucks with truckside sort for a community of 50,000 households would add between $0.47 and $3.29/household/year, depending on plastics recycled, collection bin, inclusion of perforator and the other specific details of each program. The above costs include a diversion credit of $94.00/ton. The cost of adding plastics to the three programs without diversion credit would result in an additional cost of between $0.72 and $4.32/household/year, depending on the option chosen.

A similar analysis for the HRG and Minnetonka recycling programs was also conducted. The cost of adding plastics to a 50,000 household Minnetonka type route (i.e., collecting the plastics, using the collection vehicle arrangement and using the plastic generation rates of the Minnetonka program) was estimated at $3.80/household-year, and the cost of adding plastics to a 50,000 household HRG type route was estimated at $1.42/household-year. These costs do not include diversion credit.

The Minnetonka and HRG programs, which have weekly setout rates of 16 lbs./setout and 21 lbs./setout, respectively, had curbside collection costs less than the associated waste diversion credit of $94.00/ton regardless of whether plastics were collected. By comparison, the Minneapolis A, B and C routes had average weekly setouts of 10.75 lbs./setout, and for this reason did not have collection costs less than the associated diversion credit.

In order to determine the total cost of curbside collection (for plastics as well as other recyclables), it is necessary to add the above incremental plastics collection costs to the costs associated with curbside collection if plastics were not included. Also, there will be additional cost for performing sorting, baling and granulating even though the above stated costs include curbside sortation (in comparison to commingled collection).

Table 4.3 Plastic Collection Economic Analysis for Minneapolis, MN Area [CSWS, 1990]

Parameters	No Plastics	Type of Plastic Collection		
		Route A (Soft Drink & milk bottles)	Route B (All plastic bottles)	Route C (All rigid plastic containers)
Input Parameters				
Recycling Data				
Plastic recyclables (lb/setout/wk)	0.0	0.390	0.580	0.680
Other recyclables (lb/setout/wk)	10.75	10.75	10.75	10.75
Number of households served	50,000	50,000	50,000	50,000
Plastics setout rate (%)	0	26.1	62.3	62.1
Other recyclables setout rate (%)	74.5	74.5	74.5	74.5
Cost Data				
Additional container cost ($)	0	0	5	5
Interest rate (%)	8.5	8.5	8.5	8.5
Payback period (years)	7	7	7	7
Labor cost ($/hr)	$18.50/hr	$18.50/hr	$18.50/hr	$18.50/hr
Operation & Maintenance cost ($/hr)	$5.00/hr	$5.00/hr	$5.00/hr	$5.00/hr
Administrative cost (%)	10	10	10	10
Labrie Recycling Vehicle Cost Parameters				
Recycling vehicle cost ($)	85,600	85,600	85,600	85,600
Plastics perforator cost ($)	0	10,000	10,000	10,000
Isuzu Recycling Vehicle Cost Parameters				
Recycling vehicle cost ($)	50,000	50,000	50,000	50,000
Plastics bin cost ($)	0	0	0	0
Output Parameters				
Recycling Data				
Households served	50,000	50,000	50,000	50,000
Households served/day	5,000	5,000	5,000	5,000
Plastic setouts/day	0	1,305	3,115	3,105
Setouts/day	3,725	3,725	3,725	3,725
Plastic recyclables collected (tons/year)	0	132	470	549
Other recyclables collected (tons/year)	10,411	10,411	10,411	10,411
Using Labrie Recycling Vehicle				
Number of Vehicles Required	15	15	17	17
Additional cost of plastic collection [a] ($)	$0	$23,602	$172,088	$164,590
Plastic cost/household-year ($/hh-yr)[a]	$0/hh-yr	$0.47/hh-yr	$3.44/hh-yr	$3.29/hh-yr
Using Modified Isuzu Recycling Vehicle				
Number of Vehicles Required	14	15	16	16
Additional cost of plastic collection [a] ($)	$0	$45,356	$122,490	$115,000
Plastic cost/household-year ($/hh-yr)[a]	$0/hh-yr	$0.91/hh-yr	$2.45/hh-yr	$2.30/hh-yr

a. The additional cost of plastic collection includes a landfill savings diversion credit of $94.00/ton. The additional cost of plastic collection using the Labrie truck without landfill diversion credit would be $0.72/hh-yr, $4.32/hh-yr and $4.32/hh-yr for collection scenarios A, B and C, respectively. The additional cost of plastic collection using the Isuzu truck with trailer without landfill diversion credit would be $1.16/hh-yr, $3.33/hh-yr and $3.33/hh-yr for collection scenarios A, B and C, respectively.

It should be noted that while the above study shows cost advantages of specific vehicle types over another, there are other factors not shown which are directly related to cost. For example, a recycling vehicle which requires workers to empty a residential recycling bin at/above head level (between 5 and 7 feet off the ground) into open top collection bins on the truck 300 to 500 times/day (a typical collection rate for a recycling truck) can result in a significantly increased amount of worker back and muscle injuries. For this reason, a semi-automated side loading recycling vehicle, in which the worker empties the residential recycling bin at the waist high level into truck compartments which automatically empty themselves, may actually be preferable (even though the truck costs more) when examining total potential costs.

Other Recycling Cost Estimates

Additional economic evaluation regarding the incremental cost of curbside collection of plastic with other recyclables and MSW are reported by Rankin (1988) and Temple, Barker & Sloane (1989). Some of the economic parameters used in these reports (such as tip fee and market prices) are not representative of current conditions. However, the reports concluded that there was a net economic benefit associated with adding plastic bottles to the collection program studied.

4.3 Collection Times

The time it takes to collect recyclables has a direct impact on the economics of curbside collection. Discussed below are some field measurements of collection times using varying collection methods. Assuming collection methods are similar, the collection times shown can be used for assessing curbside collection of plastics and curbside recycling in general.

The Madison, Wisconsin pilot, which collected recyclables commingled in clear plastic bags, recorded the collection times shown in Table 4.4. The average collection time for all four collection trucks utilized is about 1 minute per stop. This does not include dumping time, break time, and other time spent not related to collection, but does include time driving to and from the collection routes. No separation of materials was performed at curbside.

Collection times with and without rigid plastics recycling were recorded in the Barrhaven, Ontario pilot discussed in Chapter 2. A bin was utilized for residents to place recyclables in. The collection utilized *Labrie* type semi-automated side loading trucks

Table 4.4 Collection Times of Commingled Recyclables in Madison, Wisconsin [City of Madison, 1990]

Collection Truck Type	Average Time/Setout (minutes/stop)
Low entry side-loading refuse packer (1 person)	1.06
Low entry rear-loading refuse packer (1 person)	1.13
Conventional cab rear-loading refuse packer (2 person)	0.75
Pick-up truck with attached trailer (2 persons)	0.84

where rigid plastics, glass, metals and old newspaper were each placed in separate compartments of the truck at curbside. The time required for each setout without plastics averaged 16 seconds, of which about 3 seconds was for travel to and from the truck and 13 seconds was for truckside sorting. Bins which contained rigid plastics required an average of another 7.5 seconds to sort at the truck. These figures do not include time between residences, dumping time or time required going to and from the collection route.

Collection times were also recorded in the Minneapolis area pilot to assess the additional time required based of the plastic types collected and the vehicle type. The times recorded for each of the route/truck combinations are shown in Table 4.5. The additional time to collect plastics varied from 3-11 seconds per setout of recyclables, with clear HDPE and PET soda bottle collection requiring the least time and rigid plastic containers requiring the most time. When adjusting the time to collect plastics to only households which set plastics out, the collection time per setout is as shown in Table 4.6. The differences are not large and range from 9-15 seconds. There is also no pattern between the level of plastic collection and the time per setout. For example, two clear HDPE and PET soda bottle programs required 9 and 13 seconds, and all plastic bottle collection required 10 and 12 seconds.

The Minneapolis B and C routes, which used one container for plastics collection and one container for all other recyclables were reported to be effective. Such containers made the plastics collection easier for workers since it was not necessary to sort plastics from other container types.

The total number of stops per day by a recycling truck vary according to the amount of curbside sorting performed, the amount of sorting performed by a homeowner, the use of municipal or private collectors and the number of crew members per truck. In general, reviews of recycling routes have indicated that an average of 500 stops/day is achieved without considering any of the above factors [Glenn, 1990, 1988].

Table 4.5 Average Time Spent on Each Collection Task per Recycling Setout [Krivit, 1990]

Route	Collection Truck Type	Other Recyclables Collection [a]	Plastics Collection Time [b]	Loading Time [c]	Tagging Time [d]	Verification Time [e]	On Vehicle Time [f]	Idle Time [g]	Other Works Time [h]	Total Time
Average Time Per Recycling Setout (seconds/setout)										
Minneapolis A	Isuzu	40	5	0	1	2	13	1	1	1:02
Minneapolis B	Isuzu	43	7	0	9	5	24	2	4	1:33
Minneapolis B	Labrie	46	8	5	3	5	15	0	0	1:22
Minneapolis C	Isuzu	47	10	0	2	8	12	0	2	1:20
Minneapolis C	Labrie	38	11	4	2	5	16	10	2	1:26
Minnetonka	Lodal	30	3	0	0	0	19	0	1	0:53
Hennepin Recycling Group	Eager Beaver	27	3	0	1	1	21	1	3	0:56

a. Other Recyclables Collection Time: The time spent sorting and collecting recyclable materials other than plastics. Depending on the route, these materials may have included newspaper, corrugated cardboard, aluminum cans, steel cans or glass.

b. Plastics Collection Time: The time dedicated specifically to the collection of the corresponding routes' recyclable plastics.

c. Loading Time: The time taken to mechanically load the collection vehicle (applicable only to semi-automated side loading "Labrie" type vehicles on the Minneapolis B and C routes).

d. Tagging Time: The time used to tag either the other recyclables or the plastic materials that could not be picked up because of noncompliance with the recycling standards.

e. Verification Time: The time spent by the driver to collect household setout information.

f. On Vehicle Time: The time spent aboard and driving the collection vehicle.

g. Idle Time: Any time such as breaks taken by the driver.

h. Other Works Time: Time spent on productive activities such as cleaning spilled materials, talking to supervisor or other drivers about collection activities, transferring recyclables, and other activities related to collection work not defined above.

Table 4.6 Plastics Collection Time Per Plastic Setout [CSWS, 1990]

Route	Collection Truck Type	Plastic Collection Time (seconds/plastic-setout)
Minneapolis A	Isuzu	13
Minneapolis B	Isuzu	10
Minneapolis B	Labrie	12
Minneapolis C	Isuzu	13
Minneapolis C	Labrie	15
Minnetonka	Lodal	9
Hennepin Recycling Group	Eager Beaver	9

4.4 Recycling Truck Costs and Truck Collection Methods for Plastics

Collection vehicles are often the largest capital expense associated with curbside recycling. There are three general styles available for recycling: open top trucks, closed body trucks and trailers. Trailers, which contain bins that may be single, segmented or removable, are commonly chosen for recycling. Depending on the application, there may be disadvantages with a trailer due to maneuverability or capacity. Open top trucks have an open top and are typically loaded through a series of doors along the side of the truck, which can slide up as a compartment fills. Closed body trucks have an enclosed, partitioned collection container and are loaded from the side or top through an opening. Closed body trucks are also manufactured with semi-automated loading devices, such as a trough along the side of the vehicle, which can be emptied hydraulically into the top of the vehicle container. Appendix A provides a listing of the manufacturers of recycling vehicles and recycling trailers.

Trailers are by far the lowest cost collection option. Their expected base cost ranges from $12,000 for a 15 yd^3 trailer to $18,000 for a 22 yd^3 trailer. The base price of an open top truck will range from $25,000 to $70,000, with an average price of around $45,000/truck. Open top trucks have capacities of 15 to 25 yd^3. The price of a closed body truck ranges from $50,000 to $80,000, and semi-automated closed body trucks cost $70,000 to $90,000. Closed body trucks are the largest and have capacities of 20 to 35 yd^3. Prices of specific truck types are reviewed in Biocycle, 1989.

The following relative truck volumes have been estimated by the Center for Plastics Recycling Research for collection of recyclables: old newsprint, 23.2%; glass bottles, 13.0%; steel cans, 10.7%; aluminum cans, 16%; PET beverage bottles, 17.6%; and HDPE milk, water bottles, 19.0%. A recycle composition study with the per household generation and density conversion will help a locality determine the volumes to expect.

It is estimated that collection of uncrushed plastic HDPE and PET beverage bottles can occupy 37% of a collection truck's volume while only contributing to 5% by weight of the load. If collection of plastic fills recycling truck bins and results in modifying a collection route to clear the load prior to the filling of other truck bins (such as old newspaper), the cost of plastic collection can rise significantly. Simple solutions have been suggested:

- Add a cage on the top or back of the truck to hold plastics.
- Use netting or a bag on the side of the collection vehicle to hold plastics.
- Collect plastics in transportable bags which can be removed and replaced with empty bags when full.
- Put plastics in an unused portion of the collection vehicle.
- Add or modify the collection vehicle to include compaction.

All but the last option requires curbside sorting of materials or at least the plastic from other recyclables. However, many curbside recycling programs use a commingled collection. Adding on-board compaction is offered commercially but its effectiveness is not well known. Problems with on-board compaction are the additional cost of modifying an existing collection vehicle and the resulting minimal net volume savings after addition of the compactor. The Council for Solid Waste Solutions is conducting research on increasing compactor effectiveness. The results should be available in 1991.

One method of reducing plastic volume is to educate consumers to crush their plastic bottles prior to disposal. The National Association for Plastic Container Recovery, (NAPCOR) an industry trade group which promotes plastic recycling, provides extensive media and mailing services to assist communities in educating homeowners.

4.5 Process Cost

Manual separation of plastic bottles, the current most commonly accepted method, is estimated to sort anywhere from 1 to 6 bottles per second per sorter with a conveyor belt/manual pick station arrangement. A 1 bottle/second pick speed at an average bottle weight of 0.14 - 0.15 pounds/bottle results in a process rate of 500-550 pounds per hour. At $10/hour labor, the sort cost is $0.02/lb., not including overhead, benefits, baling, grinding, or shipment.

The general cost for plastic handling and processing has been estimated as follows [PRC, 1990b]:

- Sorting   2 - 3 ¢/lb
- Baling   3 - 4 ¢/lb
- Grinding   3 - 4 ¢/lb.
- Cleaning (flake input/output)   10 - 15 ¢/lb
- Pelletizing   5 - 7 ¢/lb

As an illustration of the overall sorting costs, Somerset County, NJ, which has mandatory recycling of any plastic bottles and manual sortation at its MRF, processed about 8,000 lb/day of plastic bottles in 1990. When the material is brought into the MRF, a negative sort is performed on the material to remove all non-bottle items and bottles that are not marketable. The end result is a mixture of PET, HDPE clear and colored, and PVC bottles which are marketed in a baled commingled state. The county receives $0.03/lb. for the commingled bales. The operation utilizes county employees at an average rate of $7.50/hour along with a work detail from the county jail. Daily manpower costs are $660, with overhead being another $200, for a total daily cost to run the plastic operation estimated at $860/day. After revenue, the cost is reduced to $620/day [Lazo, 1990]. This puts the approximate overall sort and baling cost at 11 ¢/lb, not including revenue. The Rutgers Center for Plastics Recycling Research has similarly estimated plastic bottle sorting and baling costs at about 12 ¢/lb [Dittman, 1990].

Baling equipment is typically necessary for plastic as well as paper product processing. There are two types of balers, vertical and horizontal. The baler type describes the ram movement in the baler. Horizontal balers are geared for large recycling operations and vertical balers for small recycling operations. A summary of the capability of each type is shown in Table 4.7.

4.6 *Cost Estimate Computer Programs*

Because there are so many variables, it is difficult to assess what the exact cost of recycling will be for a locality. Computer programs can be used to estimate the various costs involved in recycling.

Eastman Chemical

A computer modeling program by Eastman Chemical, a major PET producer, shows that including plastics can reduce the overall costs of a recycling program. In general, the program projects that recycling is the most cost effective when at least

Table 4.7 Recycling Center Baler Types [Firpo, 1990]

Description	Horizontal Baler	Vertical Baler
Typical Bale Size	30" x 40" x 72" 40" x 48" x 72"	30" x 48" x 60"
Bale Weight (lbs)	700 - 1,800 lbs (depending on plastic density and baler pressure)	600 - 1,000 lbs (for cardboard) 350 - 650 lbs. (for plastic)
Power requirements (Hp)	30 - 150 Hp	10 - 20 Hp
Cost ($)	$30,000 - $150,000	$8,000 - $13,000
Advantages	• Higher production than vertical (3 - 10 bales per hour) • Higher bale density than vertical baler	• Requires less than 100 square feet of total operating and material storage space • Does not require special-assembly or maintenance
Disadvantages	• Requires large floor space • May require specialized assembly or maintenance • More technical to operate • Greater safety precautions required	• Slower production than horizontal baler • Lower bale density than horizontal baler

newspaper, glass, aluminum, HDPE and PET beverage bottles and steel cans are collected from at least 100,000 households and landfill fees are over $50/ton. The cost of adding PET to commingled collection for 100,000 homes is about $148/ton and will range from $125 to 200/ton, depending on what else is collected. Adding glass is $56/ton, steel is $49/ton, newspaper is $36/ton and aluminum is $83/ton (Figure 4.1). The above values do not include revenue from the sale of recyclables. When deducting the revenue gained from the cost of adding a material, the cost of PET and/or milk bottle collection is about the same as that for other recyclables (Figure 4.2). It is also estimated that crushing plastic bottles can have a significant effect on the cost of adding plastic to a program, and even make PET a net revenue earner depending on the amount of volume reduction. For a typical community of 100,000 with recycling, the cost of MSW disposal (including recycling cost) decreases when recycling exceeds the 3% level. The computer program is available through the National Association for Plastic Container Recovery (NAPCOR). The NAPCOR address is 5024 Parkway Plaza, Charlotte, NC 28217, (704) 357-3250.

Wasteplan

The Illinois Department of Energy and Natural Resources (IDENR) also makes available a computerized integrated solid waste management planning model called WASTEPLAN (developed by Tellus Institute of Boston, MA). The program addresses composting, recycling, incineration and landfilling options. Plastics can be included in the recycling portion of the model. It has a menu driven structure and allows for variation of a variety of input data including waste stream definition, solid waste generation, recyclable material, collection systems, and processing and disposal facilities. It also is equipped with a default data file to allow for learning and customization. IDENR correspondence should be addressed to the Office of Solid Waste and Renewable Resources, 325 W. Adams Street, Springfield, IL 62704, (800) 252-8955.

Least-Cost Scheduling

A least cost scheduling of recycling has been proposed by Lund [1990a, 1990b]. Linear programming can be used to minimize the present and future costs of recycling, landfilling and waste disposal. It allows the user to address the limited capacity of an existing landfill. The method can also be used to assess the recycling decisions of waste collectors which do not greatly affect the lifetime of a landfill operated by another entity. The model accounts for the following variables: recycling option costs, closure costs, construction costs of future waste disposal facilities, revenues from recyclables, future landfill life, household generation rates and market prices. Estimates for

66 Mixed Plastics Recycling Technology

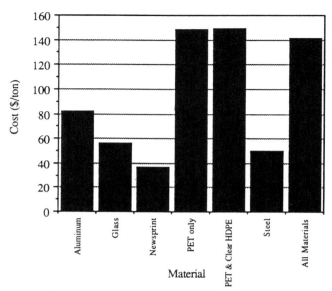

Figure 4.1 Example Costs of Including Recyclables in a Curbside Collection Program [Cornell, 1990]

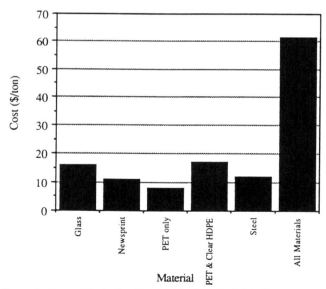

Figure 4.2 Example Costs of Including Recyclables in a Curbside Collection Program when the Revenue of Recyclables is Included [Cornell, 1990]

these variables must be made prior to entering the program. The method does not include indirect costs such as environmental impacts (e.g., aquifer contamination). However, direct costs related to environmental impacts, such as insurance coverage, may be included.

An example which includes the costs associated with landfill closure and replacement landfill construction may be used to illustrate this method. A small city (10,000 households) wishes to implement recycling to defer landfill closure and future replacement costs of waste disposal facilities. Three options are considered:

- Option 1 Recycling of household waste paper (old newspaper and junk mail) and old corrugated cardboard (OCC) with weekly collection of the separated material by a recycling vehicle. Estimated recovery rate is 70% of the waste paper and 50% of the OCC.
- Option 2 Recycling of glass, steel cans, ferrous and aluminum in addition to the household waste paper from option 1. The recovery rates are 75%, 70%, 70% and 70%, respectively. Option 2 requires a larger truck moving more slowly than the option 1 truck.
- Option 3 Collection and composting of yard waste. The estimated recovery rate is 90% of all yard waste and 30% of dirt disposed (much of the dirt comes from yard waste).

The MSW composition for the three options is shown in Table 4.8. The existing landfill has a capacity of 1,000,000 yd^3, and the cost of closure will be $8,000,000 in the year it is closed. The cost of replacing the landfill with a series of future landfills and/or incinerators covering a 50 year planning period is $82,000,000 in the year the landfill is closed. The current population is 10,000 households and expected to grow at 500 households per year.

The cost and landfill effectiveness of each recycling option is shown in Table 4.9. Since option 2 requires only a larger truck on routes used for option 1, the cost of option 2 is only the incremental cost of running trucks slower to pick up, process and market the additional material.

Without recycling, the landfill would be filled and closed in 11 years. If all recycling options were implemented for all households, there would be a 65% volume reduction in the rate of waste disposal in the landfill, and landfill life would be extended another 13 years to year 24. Figure 4.3 shows the present value of cost and savings calculated at each year of projected landfill closure beyond the initial minimum of 11 years. The least cost option would yield a landfill lifetime of 21 years. This is 10 years greater than if no recycling option were implemented, but three years shorter than if all recycling options were implemented on all households at all times. The model output also indicates the optimum implementation year of each option. Based on the cost output data, recycling options 1 and 2 would be implemented immediately and yard waste collection would not be scheduled until year 15. Extending the landfill lifetime until year 21 by recycling would result in a net present value savings of $9,1000,000.

Table 4.8 MSW Composition for Least Cost Scheduling Recycling Example [Lund, 1990a]

Material	Option [a]	Generation (lbs/hh/yr)	Landfill Volume (ft^3/yr)	Capture Rate of Recyclables [b] (%)
Paper	1,2	2,202	86.4	70
Cardboard	1,2	332	26.8	50
Glass	2	383	19.0	75
Tin cans	2	266	8.7	70
Non-ferrous metals	2	77	1.4	70
Ferrous metals	2	220	3.9	70
Garden Waste	3	731	28.1	90
Wood	3	179	3.6	90
Dirt, etc.	3	56	1.6	30

a. See text for description of options.
b. From all MSW generated.

Table 4.9 Input Parameters for Least Cost Scheduling Linear Program Example [Lund, 1990a]

	Option 1	Option 2	Option 3
Recycling Reduction (ft^3/hh/year)	73.9	24.0	28.9
Annual cost ($/hh/year)	40	15	40
Cost Effectiveness ($/ft^3/year)	0.54	0.63	1.38

People per household	4
Number of options considered	3
Rate of household growth	10,000+500 households per year growth

Waste generation rate	195.5 ft^3/hh/year
Remaining capacity	1,000,000 yd^3

Interest rate	5%
Cost of closure	$8,000,000
Cost of replacement	$82,000,000

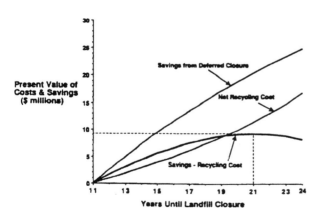

Figure 4.3 Net Present Value of Disposal and Recycling Costs for Differing Landfill Lifetimes [Lund, 1990a]

5. Markets and Packaging Changes for Recycled Plastics

5.1 Recycled Resin Demand

The demand for recycled resins is expected to rise significantly in the next three years, according to the study, "The Market for Plastics Recycling and Degradable Plastics," by Find/SVP, a New York market research firm. HDPE and PET are estimated to comprise 65% of the recycled resin market for 1990. It is expected that the amount of recycled HDPE and LDPE will nearly double between 1990 and 1991, from 252 to 498 million pounds, and from 87 to 163 million pounds, respectively [Charnas, 1990]. It is estimated the HDPE recycle will exceed 1 billion pounds by 1994. Figure 5.1 shows the recycled resin demand for the six primary thermoplastics by type from 1988 to 1993. The combined total of recycled resin for the six plastics by the end of 1993 are projected to be 3.5 times the 1990 levels.

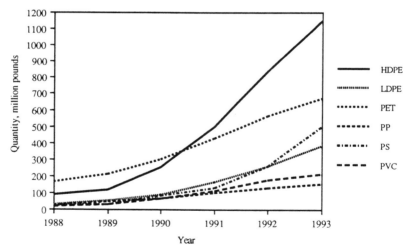

Figure 5.1 U.S. Demand of Recycled Resin, by Type, from 1988 to 1993 [Charnas, 1990]

Recycled resin prices for various stages of processing are shown in Table 5.1. The pricing information is derived from a weekly plastics publication pricing chart during middle 1990 and interviews with recyclers, resin brokers and traders. In the long term, the price of recycled resin floats relative to virgin resin prices. A common price for a baled mixture of clear and colored HDPE bottles and PET bottles which have been cleaned of foreign material is 2-4 cents per pound. Prices for off-specification, cleaned post-consumer clear HDPE pellets are at 60-70% of virgin resin prices. Cleaned, post-consumer clear HDPE flakes now float at 50% of virgin resin prices [PRC, 1990b]. It is believed that the price of recycled resin will approach that of virgin resin in the near future due to the demand from large users such as the Coca-Cola company, Procter & Gamble and Lever Brothers.

The effect of colored recycled resin on price can be seen in Table 5.1. While clean green regrind PET is selling at 17¢/lb, clear is selling for twice as much. Overall, the two major recycled resins, PET and HDPE, continue to hold or appreciate in value against

Table 5.1 Recycled Plastic Resin Prices (in cents per pound) [a] [Plastics News, 1990b]

Resin Type	Baled	Clean Regrind or Flake	Pelletized
HDPE			
Homopolymer, natural	8 - 11	28 - 31	32 - 39
Copolymer, mixed colors	5 - 8	22 - 23	28 - 31
LDPE/LLDPE Film			
Printed	5 - 9	-	19 - 23
Non-Printed:			
Opaque	7 - 12	-	-
Clear	10 - 14	-	22 - 29[b]
Filter	12 - 16	-	-
Drool	5	-	-
PET			
Clear	8 - 10	33 - 35	42 - 45
Green	7 - 8	17	-
PP Homo/Copolymer	-	8 - 20	20 - 30
PS			
Crystal	-	16 - 18	26 - 27
High Impact	-	20 - 23	30 - 33
PVC			
Blister Pack	-	8 - 10	-
Clear bottles	6 - 10	12 - 25	-

a. Prices surveyed during mid-1990.
b. Low end is for random color contamination, high end is for natural material.

the same virgin resin, and are expected to rise in the future. While the price of virgin HDPE declined in 1990, the prices for clean HDPE regrind has remained relatively constant in the range of 22-30¢/lb. The price for recycled PET increased during 1989 and 1990, and prices for PS and PVC remained constant. The prices for clean regrind are generally about half that of virgin material. LDPE, LLDPE and PS are expected to increase in value sooner than other plastics primarily because major plastics suppliers are involved in converting polyethylene and PS waste into reusable material.

Occidental Chemical Corporation has established a buy-back program for post-consumer PVC bottles. For baled material in lots of 5,000 lb and up, with 90% minimum PVC content and without caps, prices paid during the fourth quarter 1990 were as follows: clear PVC bottles with wash removable labels, 9¢/lb; mixed colors and clear PVC bottles with non-wash removable labels, 6¢/lb; clear 5 gallon PVC water bottles, cut in half and nested for shipment, 10¢/lb. Freight is paid by Occidental and truckload quantities greater than 20,000 lbs receive an additional 1¢/lb.

5.2 Packaging Changes to Increase Recycle Rates

Plastic is often not included in recycling because the costs of collection and processing do not offset the revenues gained. A thrust in plastic recycling today is to decrease the post-use processing necessary to achieve separated resins for reuse. The cost to process resins with acceptable levels of foreign plastic and non-plastic contaminants may only be marginally less than the value of the processed resin. It is desirable for collected plastics to be separated from each other so there is greater value in the material and broader application potential. As a consequence, modification of plastic packaging design methods is necessary to obtain a higher value product for recycling following consumer/industry use. Guidelines for packaging design to minimize its contribution to the solid waste stream and add more plastics to the recycle stream have been proposed [Selke, 1990]:

- *Use reusable packages*
  If a package can be reused in its original application, it can be very effective at waste reduction by eliminating disposal requirements for several cycles. Obviously other considerations must enter into this decision. If the package is not returned, it cannot be reused, so the cooperation of the user is crucial. This option is likely much easier to implement for distribution packages than for consumer ones. Costs and energy requirements of returning and cleaning containers must also be analyzed. In many cases, containers will have to be stronger to permit reuse and therefore will use more material. The net benefits must be carefully calculated.

- *Use a single material, wherever possible*
  Multimaterial packages are, in general, less suited to recycling than single material packages. All plastic containers are preferred to plastic with paper and/or aluminum. Single resin plastic containers are preferable to multi-resin plastic containers.

- *Use materials that are either easily separable or compatible if a single material cannot be used*
  If a multimaterial structure is needed, the goal is to design that structure in a way that does the least damage to recycling potential for that package. The HDPE base cup on a PET beverage bottle is not a serious problem for recycling because a relatively simple water flotation process will separate lighter than water HDPE from heavier than water PET.

- *Use recycled materials where possible*
  The existence of markets for recycled materials is a key part of any recycling operation. It does no good whatsoever to separate and process materials if they do not find uses in new products. The packaging industry has an obligation to increase its use of recycled materials. While there obviously are applications for which only virgin materials are suitable, they should not be specified unless valid reasons for the exclusion of recycled materials exist. Writing specifications based on performance rather than material content may aid in avoiding the unnecessary exclusion of recycled materials.

- *Eliminate toxic constituents*
  Packaging designs should incorporate only nonhazardous materials whenever possible. In particular, heavy metals used in additives, colorants and inks should be eliminated. If a company cannot achieve the desired color without heavy metals, perhaps a change in color coupled with a publicity campaign to let the public know the reason for the change could actually greatly enhance sales.

Modification of plastic packaging using the above methods has the potential for increasing the quantities of plastics recycled and improving the economics of the recovery process. Plastic bottle recycling can serve as one example for illustrating the potential benefits. A municipality with a recycling program may work with local plastic bottle producers to improve bottle design so that after collection, plastic bottles could be sent directly to end users without expensive processing, and more desirable resin materials such as PET and HDPE would be used. Design criteria similar to the above guidelines have been submitted to support such an effort in plastic bottle production [Anderson and Brachman, 1990]:

- Bottle products in clear rather than colored resins (with the color on the label if necessary)

- Do not use adhesives on the labels (use shrink wrap, for example) or utilize easily soluble adhesives
- Use non-aluminum caps made from the same resin and with the same viscosity as the bottle body
- Provide easily recognized labeling of plastic type
- Use industry supported campaigns to educate the public to flatten plastic containers

The first three criteria are initially subject to front end product testing to ensure safe packaging, while the remaining two criteria are tied into post-consumer use. Part of the incentive in developing future consumer plastic packaging which has higher value due to uniform resin composition is the processing which could be eliminated. This will achieve the greatest value from the recycled plastic by allowing a manufacturer which uses recycled content resin to directly purchase waste plastics without the cost of an intermediate processor. This would also allow municipalities to bypass intermediate processors and go directly to end users. This process is conducted in some municipality/company arrangements with the recycle of clear HDPE bottles which are often baled or reground by a recycling program and sold directly to a manufacturer which uses secondary plastics.

5.3 Markets in Primary Recycling

Primary recycling, or the converting of otherwise waste plastic into products similar to the original product, is generally the most favored form of recycling. It is desirable because it is not necessary to create a new market niche for a product and because the need for virgin resins are reduced. The plastic manufacturing industry regularly recovers waste "trim scrap" generated in-house because it is convenient to do so, and because the scrap material is contaminant free and of a known composition. The packaging industry is moving beyond in-house recycle of plastic scrap by manufacturing bottles containing a certain portion of post-consumer plastic products primarily because of its market appeal and recycling goals set at the federal, state, and local levels.

Plastic users have made a market for recycled plastic by modifying machinery for two and three layer coextrusion heads. Coextrusion is a method applied primarily to HDPE and sandwiches a recycled plastic layer in between virgin resins. It is used because it produces a uniform appearance of bottle exteriors and a market safe container. Coinjection stretch blow molding of PET is another fabrication method being looked at. It is capable of producing multi-layer bottles with recycle PET sandwiched in between. Examples of recycled resin in packaging are shown in Table 5.2, a majority of them being bottle coextrusion processes.

5.4 Markets in Secondary Recycling

Secondary recycling is the recycle of plastic resins into new products with less demanding physical and chemical characteristics than the original application. Mixed plastics are applied most easily in the secondary recycling market because less separation of resin types and less complicated production methods are necessary to achieve a finished product. The most readily recognized secondary plastic product is "plastic lumber" (thick extrusion molded slabs of resin in which some types of resins act as fillers). Plastic lumber is used to make park benches, fence posts, boat docks, playground equipment and the like. The fabrication of mixed plastic lumber is discussed in Part II of the book.

Examples of other value-added products which can be made (via secondary recycling) include products such as recycling containers, refuse containers, flower pots, greenhouse potting trays, traffic cones, speed bumps, downspout splash blocks, etc. Rubbermaid is producing a variety of refuse containers containing 10-25% HDPE regrind and office accessories and food service trays containing 10-50% PS regrind. Another company, Utility Plastics of Brooklyn, NY, is injection molding traffic cones and barriers containing waste HDPE.

Some recovered post-consumer and post-industrial resins are for sale in pellet form for manufacturers to utilize in production. Table 5.3 shows producers/companies which produce such pellets.

Table 5.2 Primary Recycling of Post-Consumer Plastic Resins

Company	Resin	Description
Astro Valcour	LDPE	Air bubble cushion packaging containing 20% post-consumer and 30-40% in-house scrap
Dolco Packaging	PS	Egg carton manufacture
Drug Plastics & Glass	HDPE	10 - 30% recycle material in a stock bottle made using coextrusion
Jennico	HDPE	50% recycle detergent containers made on extrusion/blow molding machines
Johnson Controls	PET	100% recycle motor oil, transmission fluid and other non-food product container; the product is stretch blow molded
Lever Brothers	HDPE	25 - 35% recycle in multiple detergent and fabric softener containers produced using coextrusion
Mobil Chemical	HDPE	Post-consumer grocery sacks used to re-manufacture grocery sacks
PCL & Eastern Packaging	LLDPE HDPE	Post-consumer grocery bag sack used to manufacture film bags
Plax	HDPE	Two and three layer household chemical and motor oil bottles
Procter & Gamble	HDPE	20 - 30% recycle in multiple detergent and fabric softener containers produced using coextrusion. Supplier companies are Plastipak Packaging, Owens-Brockway Plastic Products and Continental Plastic Containers
	PET	100% recycle in one floor cleaning product bottle
Sonoco Graham	HDPE	15 -20% recycle content in motor oil bottles
	HDPE	Post-consumer grocery sacks used to manufacture bottles, pipe
Vangaurd Plastics	HDPE	Post-consumer grocery sacks used to re-manufacture grocery sacks
American Mirrex	PVC	Rigid vinyl packaging films made 30% post-industrial recycled vinyl

Table 5.3 Suppliers of Recycled Plastic Resin

Company	Resins Available	Comments
Denton Plastics Portland, OR	HDPE PP PS ABS	(503) 257-9945
Envirothene Chin, CA	HDPE LDPE PP	Primarily HDPE available for sale. Plant start-up 1/91. (714) 465-5144
Georgia Gulf Plaquemine, LA	PVC	(504) 685-1235
Polymerland Itasca, IL	ABS Polycarbonate (PC) PC/Polybutylene-alloy	3 grades ABS available (800) 752-7842
Reprean Waterbury, CT	Flexible and rigid thermoplastics	(203) 753-5147
Secondary Polymers Detroit, MI	HDPE PET	Colored and natural HDPE flake (313) 922-7000
Soltex Polymer Houston, TX	HDPE	"Fortiflex XF855" contains 25% post-consumer HDPE
United Resource Recovery Findlay, OH	HDPE	Off-white pellets from PC milk bottles and black pellets from PC colored bottles. (419) 424-8237

Appendix A: Recycling Vehicle Equipment Manufacturers

Able Body Company
P.O. Box 891
Newark, CA 94560
(415) 796-5611

Accurate Industries
Erial & Hickstown Rds.
Erial, NJ 08081
(800) 257-7340

All Seasons Recycling
62 N. Third Street
Stroudsburg, PA 18360
(717) 424-1818

American Rolloff
P.O. Box 5757
3 Tennis Ct.
Trenton, NJ 08638
(609) 588-5400

Amertek
P.O. Box 865
Woodstock, Ontario
Canada N4S 8A3
(519) 539-7461

Amthor's Inc.
Route 22 East
Walden, NY 12586
(914) 778-5576

Automated Waste Equipment Co.
P.O. Box 5757
Trenton, NJ 08638
(609) 588-5400

Brothers Industries, Inc.
P.O. Box 190
Morris, MN 56267
(800) 950-6045

Dempster, Inc.
P.O. Box 1388
Toccoa, GA 30577
(404) 886-2327

DeVivo Industries
P.O. Box 308
Botsford, CT
(203) 270-1091

Eager Beaver Recycling Equipment
Interstate 295
Thororare, NJ 08086
(800) 257-8163

Fitzgerald Truck Equipment
7512 Lovers Lane
Cattaraugus, NY 14719
(716) 257-3494

Frink America, Inc.
205 Webb Street
Clayton, NY 13624
(315) 686-5531

Full Circle
P.O. Box 250
Paris, IL 61944-0250
(217) 465-6414

Galbreath, Inc.
P.O. Box 220
Winamac, IN 46996
(219) 946-6631

Heil Co.
P.O. Box 8676
Chattanooga, TN 37411
(615) 899-9100

Holden Trailer Sales, Inc.
Route 1, Box 151
Southwest City, MO 64863
(417) 762-3218

Kann Manufacturing
414 N. 3rd. Street, Box 400
Guttenburg, IA 52052
(319) 252-2035

Impact Products
281 E. Haven
New Lenox, IL 60451
(815) 485-1808

Jeager Industries, Inc.
43 Gaylord Road
St. Thomas, Ontario
Canada N5P 3S1
(519) 631-5100

Labrie Ltd.
302 Rue de Fleuve
Beaumont, Quebec
Canada G0R 1C0
(418) 837-3606

Leach Company
P.O. Box 2608
Oshkosh, WI 54903
(414) 231-2770

Lodal Inc.
P.O. Box 2315
Kingsford, MI
(906) 779-1700

Marathon Equipment Company
P.O. Box 290
Clearfield, PA 16830
(800) 526-8997

May Manufacturing
5400 Marshall Street
Arvada, CO 80002
(800) 292-7968

Mobile Equipment Company
3610 Gilmore Avenue
Bakersfield, CA 93308
(805) 327-8476

Multitek Inc.
P.O. Box 170
Prentice, WI 54556
(715) 428-2000

National Recycling Equipment Co.
190 Service Avenue
P.O. Box 6788
Warwick, NJ 02887
(401) 738-7225

Norcia Corp.
RD 4, Box 451
North Brunswick, NJ 08902
(210) 297-1101

Parker Industries
P.O. Box 157
Silver Lake, IN 46982
(800) 526-8997

Appendix A: Recycling Vehicle Equipment Manufacturers

Peabody Galion
500 Sherman
P.O. Box 607
Galion, OH 44833
(419) 468-2120

Peerless
P.O. Box 447
Tualatin, OR 97062
(800) 331-3321

Perkins Manufacturing Co.
3220 W. 31st. Street
Chicago, IL 60623
(312) 927-0200

Peterson Lightning Cyclers
434 U.S. Hwy #27 North
Lake Wales, FL 33853
(813) 676-9688

P-M/Rudco
114 East Oak Road
Vineland, NJ 08360
(609) 696-3423

Rand Systems
P.O. Box 27746
Raleigh, NC 27611
(800) 543-7263

Refuse Trucks Inc.
4849 Murietta Street
Chino, CA 91710
(714) 590-0200

SAC Recycling
P.O. Box 769
South Windsor, CT 06074
(203) 282-8282

Scranton Manufacturing Co.
P.O. Box 336
Scranton, IA 51462
(800) 831-1858

Summit Trailer Sales
1 Summit Plaza
Summit Station, PA 17979
(717) 754-3511

Sunnyvale Truck Equipment
755 North Mathilda
Sunnyvale, CA 94086
(408) 739-5475

Tafco Equipment Co.
Hwy. 16, P.O. Box 339
Blue Earth, MN 56013
(507) 526-7346

Walinga Inc.
Rural Route 3
45 Kerr Crescent
Guelph, Ontario
Canada N1H 6H9
(519) 763-7000

WASP Recycling Equipment
Box 100
Glenwood, MN 56334
(612) 634-5126

Wayne Engineering Corp.
2412 W. 27th Street
P.O. Box 648
Cedar Falls, IA 50613
(319) 266-1712

Appendix B: Glossary

ABS (Acrylonitrile-Butadiene-Styrene) A family of thermoplastics based on these three compounds. ABS resins are rigid, hard, tough, and not brittle. This family of plastics is used to produce durable goods products such as appliances and automotive parts.

Acrylic A family of resins formed from methacrylic acid and known for their optical clarity. Widely used in lighting fixtures because they are slow burning or may be made self extinguishing.

Blow-Molding A method of fabrication in which a parison is forced into the shape of a mold cavity by interval gas pressure.

Coextrusion The process of extruding two or more materials through a single die so that the material bonds together at the mating surface.

Copolymer Typically a polymer of two chemically distinct monomers.

EPM/EPDM (Ethylene Propylene Rubbers) A group of elastomers (rubber-like material) obtained by copolymerization of ethylene and propylene for EPM and a third monomer (diene) for EPDM. Their properties are similar to those of rubber.

EVA (Ethylene-Vinyl Acetate Copolymer) Copolymers of major amounts of ethylene with minor amounts of vinyl acetate, that retain many of the properties of polyethylene but have considerably increased flexibility, elongation and impact resistance. EVA is used as an adhesive for bonding base cups to PET beverage bottles and labels to bottles. EVA is a form of LDPE

Future Value Lump sum cash value at the end of a given time period

Gaylord A container for holding waste plastic, plastic flake or plastic pellets. Often times a gaylord is a cardboard box measuring 34"x43"x38".

HDPE (High Density Polyethylene) Polyethylene plastic having a density typically between 0.940 and 0.960 g/cm^3. While LDPE chains are branched and linked in a random fashion, HDPE chains are linked in longer chains and have fewer side branches. The result is a more rigid material with greater strength, hardness, chemical resistance and a higher melting point than LDPE.

Industrial Scrap Plastic material originating from a variety of in-plant operations and which may consist of a single material or a blend of a known composition.

LDPE (Low Density Polyethylene) Polyethylene plastic having a density typically between 0.910 and 0.925 g/cm^3. The ethylene molecules are linked in random fashion, with the main chains of the polymer having long and short side branches. The branches prevent the formation of a closely knit pattern, which results in a soft, flexible and tough material.

LLDPE (Linear Low Density Polyethylene) LLDPE is manufactured at much lower pressures and temperatures than LDPE. LLDPE has long molecular chains without the long chain side branches of LDPE, but with the short chain side branches.

Appendix B: Glossary

Mixed Plastic A mixture of plastics, the components of which may have widely differing properties.

Monomer A compound which typically contains carbon and is of a low molecular weight (compared to the molecular weight of plastics), which can react to form a polymer by combination with itself or with other similar compounds.

Nylon A generic name for a family of resins which have a recurring amide groups (-CO-NH-) as an integral part of the main polymer chain. Nylons are identified by denoting the number of carbon atoms in the polymer chains of each of the constituent compounds which formed the resin. For example, nylon 6,6 refer to the number of carbon atoms in each of the two compounds used to form it.

Participation Rate The ratio of residences which set out recyclables at least once in a 4 week period to the total number of residences on the colleciton area.

PBT (Polybutylene Terephthalate) Similar to PET, but formed using butanediol rather than ethylene glycol (as with PET). PET and PBT are the two thermoplastic polyesters that have the greatest use.

PET (Polyethylene Terephthalate) A saturated thermoplastic polyester formed by condensing ethylene glycol and terephthalic acid. It is extremely wear and chemical resistant and dimensionally stable. It also has a low gas permeability in comparison to HDPE, LDPE, PP and PVC, which is why it is used so extensively for carbonated beverage bottles.

Phenolics A family of thermosetting resins made by reacting a phenol with an aldehyde. Phenolics are known for good mechanical properties and high resistance to temperature.

Polyesters A family of resins also known as alkyds. The main polymer backbone is formed through the condensation of polyfunctional alcohols and acids. Polyesters can be saturated (elements or compounds cannot be added to the main backbone) or unsaturated. One of the most important polyester is PET, a saturated polyester.

PP (Polypropylene) A thermoplastic resin made by polymerizing propylene with suitable catalysts. Its density of approximately 0.90 g/cm^3 is among the lowest of all plastics.

Present Value Lump sum cash value at the beginning of a given time period

Primary Recycling The processing of waste into a product with characteristics similar to those of the original product.

PS (Polystyrene) Polymers of styrene (vinyl benzene). PS is somewhat brittle and is often copolymerized or blended with other materials to obtain desired properties. HIPS (high impact PS) is made by adding rubber or butadiene copolymers. Commonly known PS foams are produced by incorporating a blowing agent during the polymerization process or injecting a volatile liquid into molten PS in an extruder.

PUR (Polyurethanes) A large family of resins based on the reaction of isocyanate with compounds containing a hydroxyl group. PUR can be made into foam or resin, rigid or flexible, thermoset or thermoplastic.

PVC (Polyvinyl Chloride) PVC is produced by polymerization of vinyl chloride monomer with peroxide catalysts. The pure polymer is hard and brittle, but becomes soft and flexible with the addition of plasticizers.

Recycled Plastic Plastic products or parts of a product that have been reground for sale or use to a second party, or plastics composed of post-consumer material or recovered material only (which may or may not have been processed).

Regrind Plastic Plastic products or parts of a product that have been reclaimed by shredding and granulating for use (primarily intended as an in-house term).

Resin A term which is generally used to designate a polymer, a basic material for plastic products. It is somewhat synonymously used with "plastic," but "Resin" (and polymer) most often denotes a polymerized material, while "plastic" refers to a resin which also includes additives such as plasticizers, fire retardants, fillers or other compounds.

Secondary Recycling The processing of waste into materials which have characteristics less demanding than those of the original plastic product.

Setout A setout is defined as a residence or a dwelling "setting out" its recycling container for collection. Setout rate is the ratio of residences which set out recyclables each collection period (such as weekly).

Thermoplastic Plastic that can be repeatedly softened by heating and hardened by cooling through a temperature range characteristic of the plastic, and that in the softened state can be shaped by flow into articles by molding or extrusion.

Thermoset Plastic that, after having been cured by heat or other means, is substantially infusible and insoluble. Cross-linking between molecular chains of the polymer prevent thermosets from being melted and resolidified.

References

Ackerman, K. Bottle Coding Laws. Plastics News 29(2):14, 1990.

Adams, S. Fitchburg Recycling Program Management and Expansion - Final Report. Wisconsin Department of Administration Waste to Energy and Recycling Grant Program. Madison, WI. 1990a.

Adams, S. Residential Polystyrene Recycling. Resource Recycling 9(10):72-790, 1990b.

Adams, S. Recycling Coordinator, City of Fitchburg, WI. Personal Communication. 1990.

Anderson, P., and Brachman, S. Making Plastics Recycling Practical: New Roles for Cities and Industry. Environmental Decisions 2(5):14-24, 1990.

Biocycle. CollectionVehicles Hit the Road. Biocycle 30(6):37-39. 1989.

Bond, B. "Recycling Plastics in Akron, Ohio" Proceedings of the Society of Plastic Engineers Regional Technical Conference - Recycling Technology of the 90's. Chicago, IL. 1990.

Brewer, G. Plastics Recycling Action Plan for Massachusetts. Massachusetts Department of Environmental Quality Engineering. Boston, MA. 1988.

Charnas D. Short Supply of Recycled Resins to Keep Prices Up. Plastic News September 17, 1990. 2(29):12-13, 1990.

City of Chicago. DRAFT - Solid Waste Management Plan Waste Characterization Report for Chicago, Illinois. HDR Engineering, Inc. Omaha, NE. 1990.

City of Madison. Final Report - City of Madison, Wisconsin Pilot Curbside Recycling Program. Wisconsin Department of Administration Waste to Energy and Recycling Grant Program. Madison, WI. 1990.

City of Milwaukee. Household Recycling Program - Final Report. Wisconsin Department of Administration Waste to Energy and Recycling Grant Program. Madison, WI. 1990.

City of Seattle. Pilot Program in Plastics Recycling - Final Report. Seattle Engineering Department, Solid Waste Utility Division. Seattle, WA. 1989.

CNT (Center for Neighborhood Technology). Large Scale, Multi-Resin Plastics Recycling in Chicago. Prepared for Amoco Chemical Company, Chicago, IL. 1990.

COPPE (Council on Plastics and Packaging in the Environment). Fact Sheet. Washington, D.C. No date.

Cornell, D. "Computer Model for Comprehensive Curbside Recycling Collection" Proceedings of the Society of Plastic Engineers Regional Technical Conference - Recycling Technology of the 90's. Chicago, IL. 1990.

CSWS (Council for Solid Waste Solutions). *Plastics Recycling Pilot Program Final Report: Findings and Conclusions - Prepared for Hennepin County [Minnesota]*. Washington, D.C. 1990.

Dittman, F. "New Developments in the Processing of Recycled Plastics" *Proceedings of the Society of Plastic Engineers Regional Technical Conference - Recycling Technology of the 90's*. Chicago, IL. 1990.

Engelbart, M. Recycling Coordinator, City of Milwaukee, WI. Personal Communication. 1990.

Eyring, W. *Mixed Resin Post-Consumer Plastics Processing Facility: A Notice of Opportunity*. Center for Neighborhood Technology. Chicago, IL. 1990.

Fearncombe, J. *Guide for Recyclers of Plastic Packaging in Illinois*. Prepared by Bottom Line Consulting, Inc. for Illinois Department of Energy and Natural Resources. Springfield, IL. 1990.

Firpo, S. "Plastic Processing at the Recycling Center" *Presentation at the Ninth National Recycling Conference*, San Diego, CA. 1990.

Franklin Associates. *Characterization of Plastic Products in Municipal Solid Waste - Final Report*. Prepared for Council for Solid Waste Solutions. Prairie Village, KS. 1990.

Glenn, J. Special Report: Curbside Strategies-Junior, Take Out the Recyclables. *Biocycle* 29(5):26-31. 1988.

Glenn, J. Curbside Recycling Reaches 40 Million. *Biocycle* 31(7):30-37. 1990.

Hanesian, D., Merriam, C., Pappas, J., Roche, E., Rankin, S., and Bellinger, M. Post-Consumer Plastic Collection - A Source of New Raw Material From Municipal Solid Waste. *Journal of Resource Management and Technology* 18(1):35-39, 1990.

Hegberg, B., Hallenbeck, W., and Brenniman, G. *Technologies for Recycling Post-Consumer Mixed Plastics*. (OTT-8) University of Illinois at Chicago, Office of Technology Transfer. Chicago, IL. 1991. (Part II of this book.)

Hill, J. "Rubbermaid wants Higher Quality Recycled Resin from Plastic Suppliers" *Recycling Times* 2(12):3, 1990.

Krivit, D. Consultant to CSWS Minneapolis, MN Curbside Collection Project [CSWS, 1990]. Personal Communication. 1990.

Lazo, A. Recycling Coordinator for Somerset Co., Somerville, NJ. Personal Communication. 1990.

Lund, J. Least Cost Scheduling of Solid Waste Recycling. *ASCE Journal of Environmental Engineering* 116(1):182-197, 1990a.

Lund, J. Economic Analysis of Recycling for Small Municipal Waste Collectors. *Journal of Resource Management and Technology* 18(2):84-96, 1990b.

Madison, City of *Final Report - Pilot Curbside Recycling Program*. Wisconsin Energy Bureau. Madison, WI. 1990.

Modern Plastics. U.S. Resin Sales. <u>Modern Plastics</u> 67(1):99-109, 1990.

Moore, P. "An Overview of Collection Programs for Post-Consumer Plastic" <u>Presentation at the Ninth National Recycling Congress</u>. San Diego, CA. 1990.

Morrow, D., and Merriam, C. "Recycling - Collection Systems for Plastics in Municipal Solid Wastes - A New Raw Material" <u>Proceedings of Society of Plastics Engineers RETEC Conference - New Developments in Plastics Recycling</u>. Charlotte, NC. 1989.

Peritz, L. Vice President, wTe Corporation. "Plastic Processing Technology Case Study" <u>Presentation at the 1990 National Recycling Congress</u>. San Diego, CA. 1990.

Plastics News. PET Recycling Hits All Time High in 1989. <u>Plastics News</u> November 5, 1990. 2(36):2, 1990a.

Plastics News. Resin Pricing Chart as of Nov. 1, 1990. <u>Plastics News</u> November 5, 1990. 2(36):33-35, 1990b.

PRC. Post-Consumer Plastics: Densities, Weights and Sizes. <u>Plastics Recycling Compendium</u> Information Resource No. 111/A. 1990a.

PRC. Recycled HDPE Resin Costs and Pricing. <u>Plastics Recycling Compendium</u> Information Resource No. 115/A. 1990b.

Rankin, S., Ed., Frankel, H., Hanesian, D., Merriam, C., Nosker, T. and Roche, E. <u>Plastics Collection and Sorting: Including Plastics in a Multi-Material Recycling Program for Non-Rural Single Family Homes</u>. Center for Plastics Recycling Research, Rutgers, The State University of New Jersey, Pistacaway, NJ. 1988.

Rankin, S. Recycling Plastic in Municipal Solid Wastes. <u>Journal of Resource Management and Technology</u> 17(3):143-148, 1989.

Schut, J. A Barrage of News From the Recycling Front. <u>Plastics Technology</u> 36(7):109-119, 1990.

Selke, S. <u>Packaging and the Environment</u>. Technomic Publishing. Lancaster, PA. 1990.

TBS (Temple, Barker and Sloane). <u>The Incremental Costs of Adding Plastic Bottles to Curbside Recycling in New Jersey: Five Case Studies</u>. Prepared for The Plastic Recycling Corporation of New Jersey. Washington, D.C. 1989.

TransOntario Plastics Recovery, Inc. <u>Barrhaven [Ontario] Demonstration Project: Collection of Rigid Plastic Containers in the Blue Box</u>. Prepared for Ontario Ministry of the Environment and Society of the Plastics Industry of Canada. Environment and Plastics Institute of Canada. Don Mills, Ontario, CA. 1989.

U.S. EPA Office of Solid Waste. <u>Methods to Manage and Control Plastic Wastes - Report to Congress</u>. Publication Number EPA/530-SW-89-051. Washington, D.C. 1990a.

U.S. Environmental Protection Agency (EPA) Office of Solid Waste. <u>Characterization of Municipal Solid Waste in the United States: 1990 Update</u>. Publication Number EPA/530-SW-90-042A. Washington, D.C. 1990b.

Watson, T. Products from Plastics - Will the Profits Follow ? <u>Resource Recycling</u>. 9(7):50-56, 1990.

Part II

Recycling Technology

The information in Part II is from *Technologies for Recycling Post-Consumer Mixed Plastics—Plastic Lumber Production, Emerging Separation Technologies, Waste Plastic Handlers and Equipment Manufacturers,* prepared by Bruce A. Hegberg, Gary R. Brenniman, and William H. Hallenbeck of the University of Illinois Center for Solid Waste Management and Research for the Illinois Department of Energy and Natural Resources, Office of Solid Waste and Renewable Resources, March 1991.

Acknowledgments

This public service report is a result of the concern of the Illinois Governor, State Legislature, and the Public for the magnitude of the solid waste problem in Illinois. The concern led to the passage of the Illinois Solid Waste Management Act of 1986. One result of this Act was the creation of the University of Illinois Center for Solid Waste Management and Research. The Office of Technology Transfer (OTT) is part of this Center. One of OTT's means of transferring technology is the publication of public service reports which contain discussions of important topics in solid waste management.

Funding for this public service report was provided by the Illinois Department of Energy and Natural Resources (IDENR), Office of Solid Waste and Renewable Resources. The views expressed in this report do not necessarily reflect the policy of the IDENR. Additionally, OTT would like to acknowledge the review provided by IDENR.

Summary

Broad scale recycling of post-consumer plastic waste is technically difficult because of the variety of plastic resins which exist and the difficulty of sorting them. While further work in processing and separating waste plastics is necessary for widespread plastics recycling, there are methods to utilize mixed plastic waste and methods to clean and separate some types of plastics. The latter is primarily an emerging field of research in recycling technologies. The purpose of this report is to discuss technologies which have been developed for the separation and processing of mixed plastic wastes.

Although there are many types of plastics, eight types comprised 76% of the 1989 U.S. sales (including export sales) of all plastics: low density polyethylene (LDPE), polyvinyl chloride (PVC), high density polyethylene (HDPE), polypropylene (PP), polystyrene (PS), polyurethane (PUR), phenolic, and polyethylene terephthalate (PET). Efforts in recycling have concentrated on the thermoplastics HDPE, LDPE, PP, PS, PVC and PET because thermoplastics can be repeatedly softened by temperature increases and hardened by temperature decreases. These thermoplastics are also referred to as commodity resins because they are produced in the largest volumes at the lowest cost and have common characteristics among producers. Consumption of these six thermoplastics is led by the packaging industry. PUR can be formed as a thermoplastic or a thermoset (a resin which has undergone a chemical reaction leading to a relatively infusible state that cannot be reformed). Phenolics are a family of thermosetting resins.

There are numerous manufacturing methods used in the production of plastic goods. Post-consumer thermoplastic products are typically made using an extrusion mold, blow mold or injection mold process. An extruder consists of a rotating screw in a barrel to melt plastic pellets and force the molten resin out the end through a die. Extruders typically precede blow mold and injection mold dies. Extruders are also used directly to form film plastics (such as dry cleaning bags or sheet wrap) or profile extrusions (such as pipe or window framing). Blow molding, which is how bottles are formed, involves the insertion of a tube of resin melt from an extruder into a mold and then blowing on the inside of the tube with a gas to form the melt against the inside walls of the mold. Injection molding is used to inject resin melt into a mold from an extruder barrel. It forms solid parts (such as lids or caps), allows precise definition (such as sharp corners) and permits close tolerances.

Plastic manufacturing involves the use of compatibilizers. Compatibilizers are types of plastic additives that can have a direct affect on the recycling of plastic mixtures. They allow for the bonding of two otherwise unadhering plastics when blended together.

This can allow for higher end use of a plastic mixture than without a compatibilizer additive. Compatibilizers exist for polyethylene (PE)/PVC blends, PE/PS blends, PVC/PS blends and other more specialized plastics.

The manufacture of flow molded linear profiles, or plastic lumber as it is commonly referred to, has received a great deal of attention as a solution to using mixed plastics. It is a method to utilize plastic containers and films en masse which could not otherwise be collected in quantities significant enough to justify separation. Plastic lumber is also a method to utilize "tailings," the miscellaneous plastic resins left after a recycle stream has been "mined" of higher value HDPE and PET bottles. Tailings may also typically be the plastics collected by recycling that were not specified by the collecting agency. Although, from a polymer science point of view, such a diverse combination of plastics are not considered to be readily capable of "blending" into a compatible product, the mixture can easily be processed by extrusion into large cross-section items that have significant strength and utility.

Although the manufacture of plastic lumber from mixed plastics without separation has large potential as a solution to mixed plastics, there are associated problems. Depending on market prices and proximity to the manufacturer, it may be necessary to pay a manufacturer to take the waste plastic (there are no mixed plastic lumber producers in Illinois). The cost of shipping can have a large impact on the recycling operation economics. It may also be necessary to separate plastics to obtain a desired color or appearance of the finished lumber product or to attain a product with reasonable quality standards. While dark browns, blacks and grays are possible with mixed plastic bales, lighter colors such as blue, yellow and light gray are not possible without using separated clear and white HDPE/LDPE. A large proportion of LDPE, both granulated and molded, produces articles which are very elastic. Similarly, a large proportion of PP will produce articles which are brittle. Consequently, blending of granulated material by polymer may be important depending on the product to be manufactured. If separated with enough quality control, the separated resins will bring a better price than through the plastic lumber market. Finally, the manufacturer may require that a municipality collecting mixed plastics buy the product following recycling.

There are approximately nine companies producing lumber from mixed plastics collection programs, with three in the midwest area. While most companies possess a proprietary machine or license, the basic principles of each machine are the same. Post-consumer bottles or rigid plastics must first be ground and films must be densified and ground. The material is fed into an extruder barrel, forced through a die and cooled. Additives may be provided in the feed to enhance properties or set the color. LDPE,

HPDE, PP and PS all work well with the process. PVC can be used with additives. Higher melting point plastics such as PET become a filler if used in plastic lumber production. The physical properties of plastic lumber have been shown to improve significantly with the addition of PS to mixtures.

Because the manual sorting of heterogeneous plastic mixtures is the current level of technology in all but a few applications of plastics recycling, methods to effectively separate them on a more automated basis are being attempted. Methods for sorting rely on responses to differing environments such as specific gravity changes, x-ray diffraction, optical recognition and dissolution in solvents. Each of these methods is capable of separating types of plastic to a certain extent and therefore any one process cannot individually sort any type of plastic mixture. These methods can be classified into macro, micro or molecular scale sorting. Macro sorting involves separating plastic based on an entire product, such as using optical sensing to separate whole bottles by color. Manual separation is also a macro sorting method. Micro sorting involves the initial processing to a uniform criteria, such as size, with subsequent separation. An example of micro sorting is the commercially viable PET bottle/HDPE base cup separation method, where bottles are ground up and sent through a hydrocyclone; the PET and aluminum cap grind sinks while the HDPE floats. Molecular separation involves processing plastic by dissolving the plastic and then separating plastics based on temperature.

Methods to separate clear glass and PET plastic bottles from colored glass and PET plastic bottles are being evaluated at one university and a recycling center in Illinois. It involves the use of an optical sensor activating a diversion device to assist in the separation of a majority of clear bottles from colored bottles. Early testing indicated separation effectiveness was high. The device is estimated to cost about $2,000 to $4,000 for a small unit which would process 60 tons/year PET and 1,000 tons/year glass and be used less than 200 hours annually. Processing ten times as much would make better use of operating time and result in a payback of less than one year.

The separation of PVC bottles from other plastic in sorting programs has become important due to the adverse effects of even small amounts of PVC present in other plastic (e.g., PET). Upon being melted with PET, which is what PVC is most often mistaken to be, hydrochloric acid can form and corrode the metal parts used in plastic extrusion machines. A study of PVC bottles in a mixed plastic recycling program showed that manual separation is about 80% accurate in identifying which bottles are PVC. Research is being conducted to provide automated separation methods for PVC bottles. Most use x-ray fluorescence to detect characteristic backscattering from chlorine atoms in PVC which is higher than that detected by polyolefin plastic without chlorine. Mechanical separation can then occur. There are still some problems because the chlorine x-ray is weak, it does not penetrate paper labels and there is a rapid decrease in intensity as a bottle is moved away from the detector. It is projected that a minimum of 10 bottles/second could be analyzed with multiple detections per bottle. Even with automated x-ray separation, additional cleaning is necessary, and a method to perform cleaning and foreign plastic separation using a 1.35 and 1.30 specific gravity calcium nitrate solution is being studied.

The Center for Plastics Recycling Research (CPRR) at Rutgers University has developed a process for the reclamation of PET soda bottles that can be transferred to the public to increase recycling of such bottles. This was done because there are a number of companies in the U.S. which recycle PET soda bottles, but their methods are generally proprietary and not licensed out. The license and technology are available for a fee of $3,000 which includes a technology transfer manual with detailed equipment and process description, process economic estimates, safety and health parameters and quality control requirements and measurements. The Rutgers Beverage Bottle Reclamation Process is not meant to separate PET soda bottle colors or other type plastics, but rather to provide clean PET flake separated from caps, labels, base cups and adhesives. The process also provides clean polyolefin flakes which are mainly the HDPE base cup and PP caps, and aluminum chips from caps. A detailed cost estimate was performed by CPRR for a 20 million pound per year facility in a leased building which was based on attaining 20¢/lb for PE flake and 34¢/lb for aluminum. Plant operation was based on 24 hours/day and 330 days/year. A return on investment of 34%/yr at a PET flake price of 31¢/lb has been estimated.

Selective dissolution involves the separation of mixed plastics on a molecular scale by dissolving resin mixtures in a solvent. This method is currently being studied at a laboratory level. There are two methods being approached in the dissolution process. The first method uses one solvent to dissolve all resin types and the second method uses one solvent to dissolve one particular type of resin, but not others. Both methods have received attention because the plastic stream can be heterogeneous in nature and contaminants such as metals, glass, cellulose and some pigments can be removed. Selective dissolution can allow for microdispersion of polymer combinations, thereby rendering innocuous certain plastic components that may lead to manufacturing difficulties or poor physical properties. Promising solubility and recovery has been achieved thus far.

Recovered plastic can be marketed for reuse by directly dealing with a company which uses waste plastic in manufacturing, by directly dealing with a plastic processor which will buy waste plastic and market the cleaned and decontaminated product, by listing the recovered waste plastic in a waste exchange for marketing or by marketing the recovered waste plastic through a scrap resin broker. A resin broker or a plastic scrap handler is typically where plastic scrap is marketed after being collected at the post consumer, post commercial, or industrial scrap level. Shipments in truckload quantities are typically preferred, but smaller loads down to bales are usually accepted with an accompanying reduction in price paid. There are approximately 14 resin brokers or scrap handlers in Illinois with 11 in the Chicago metropolitan area, 2 in Joliet and 1 in Decatur. Five of the companies in metro Chicago are strictly brokers of plastic scrap. There are 20 additional brokers/processors in states neighboring Illinois: 8 in Wisconsin, 7 in Michigan, 2 in Iowa, 1 in Indiana, 1 in Kentucky and 1 in Missouri. Waste exchanges are typically sponsored by a state and provide a waste listing free of charge. Because of the extensive number of sources and types of plastics in waste, waste exchanges should generally be utilized only after other marketing methods (e.g., scrap resin brokers, plastic recycling companies) have been tried. There are 17 such waste exchanges in the U.S. and Canada; one is sponsored by the state of Illinois.

Specifications exist for non-plastic contaminants and other plastic contaminants in waste plastic loads. As may be expected, higher prices are paid for material with lesser amounts of contamination. Limits for non-plastic contaminants are typically no metals, less than 0.005% to less than 3% (by weight) non-plastic, and the material must be clean. Limits for plastic contaminants are typically less than 1% to less than 5% (by weight) other plastic, less than 1% (by weight) color on clear/natural bottle loads, no motor or vegetable oil bottles, and no PVC bottles. One method of addressing plastic contamination in general is to link recycle product prices to product quality on a commonly accepted standardized system such as the American Society for Testing Materials (ASTM). The ASTM D-20 Committee, which addresses plastic recycling and degradable plastics, is developing standards regarding waste plastic contamination and recycling.

1. Introduction

1.1 Plastics in Municipal Solid Waste

Recycling of plastic discards is one method of reducing municipal solid waste. They are beginning to join glass, steel, aluminum and paper as waste stream components that have been accepted into recycling programs across the country. It is difficult, however, to expand plastics recycling because of the variety of plastic wastes, the difficulty of sorting different types of plastics, the low density of post-consumer plastics wastes in comparison to other recyclables and the limited history of plastics recycling. Because of its heterogeneous nature and the amount of contaminants present, separation of post-consumer mixed plastic waste is the most difficult. Waste plastic from industrial operations are cleaner and more homogeneous in resin type and scrap form. The term "mixed plastics," a mixture of plastic types or a mixture of package/product types which may or may not be the same plastic type or color category, has been used to describe broad scale processing of post-consumer plastic waste. Mixed plastics also includes products which may be the same resin type but which have been fabricated using differing manufacturing techniques.

While it is possible to market recycled mixed plastic waste with limited separation, greater value and broader applications are achieved with homogeneous resins. Although it is possible to mix different type polymers together, the resulting physical properties are less desirable than the original components. Technological research regarding large scale separation of mixed plastic waste streams is being conducted. The advances in plastic separation technology are discussed in this report.

The 1989 domestic consumption of all plastics totaled 53.5 billion pounds, with 44.2 billion pounds, or 83%, being eight plastic types: low density polyethylene (LDPE), polyvinyl chloride (PVC), high density polyethylene (HDPE), polypropylene (PP), polystyrene (PS), polyurethane (PUR), phenolic, and polyethylene terephthalate (PET) [Modern Plastics, 1990a]. HDPE, LDPE, PP, PS, PVC and PET are thermoplastics, capable of being repeatedly softened by increases in temperature and hardened by decreases in temperature. They are also referred to as commodity resins because they are produced in the largest volumes at the lowest cost and have common characteristics among producers. Consumption of these six thermoplastics is led by the packaging industry. Polyurethane can be formed as a thermoplastic or a thermoset. Thermosets are resins which have undergone a chemical reaction leading to a relatively infusible state that cannot be reformed. Phenolics are another family of thermosetting resins.

Introduction

Although somewhat synonomous with "resin" and "polymer," the term "plastic" refers to a resin which includes additives for the purpose of providing a manufactured product. "Resins" (or polymers) are the basic materials for plastic products, and most often denote a polymerized material without consideration of specific additive.

A study of the 15 leading resins based on 1988 production identified that 44 billion pounds were disposed of in some manner [Franklin Associates, 1990]. Of all resin types produced, 29 billion pounds are disposed in the municipal solid waste (MSW) stream each year and only 1.1% of the waste plastic stream is recovered [U.S. EPA, 1990a]. The remainder is disposed as incinerator residue, sludge, industrial, or construction/demolition waste. In 1988, plastics comprised 9.2% by weight and 19.9% by volume of material discarded in MSW [U.S. EPA, 1990b]. One specific plastic type, PET beverage bottles, which have been targeted for recycling through curbside collection and container deposit legislation, has reached notable recycle rates of 23% in 1988 and 28% (175 million pounds) in 1989 [Plastic News, 1990a]. The increased cost of landfilling waste, the volume occupied by disposed plastic products, the value of the plastic waste material, as well as the mandate of 25% recycling of solid waste set by the state of Illinois make the addition of plastics to recycling programs a necessity. Increasing the recycle of plastic containers, film, and packaging in general from the waste stream is a logical next step in increasing recycle rates.

The numeric coding of six popular types of plastic by the Society of the Plastics Industry (PET - 1, HDPE - 2, PVC - 3, LDPE - 4, PP - 5, and PS - 6) with subsequent acceptance of such number coding of containers into law by a number of states across the country has made post-consumer plastics recycling more prevelant than ever. Municipalities and cities are beginning to collect plastic bottles, any type of rigid plastic container, and in some cases plastic films. Film is often the most predominant component of plastic on a weight basis. Further development of process and separation technologies is necessary for mixed plastics if widespread recycling of plastic bottles, containers and film is to be increased. This report discusses technologies which have been developed for the separation and processing of waste plastics.

Parameters of curbside plastics recycling such as the characterization, collection and costs are discussed in Part I of the book.

The following section 1.2 is an overview of plastic resin production and the primary methods utilized in manufacturing consumer goods and recycling waste plastics into new products. Also discussed are factors influencing the recyclability of waste plastic mixtures and compatibility between polymer types.

1.2 Plastic Resin Production and Product Manufacture

Production of plastic goods involves three primary steps:

- Resin manufacture by reacting oil or natural gas products into solid molecules
- Incorporation of additives to alter physical or aesthetic characteristics, or to permit processing of the resin
- Product manufacture, typically from pellet form for thermoplastics and liquid form for thermosets

Resin Manufacture

Different resins are manufactured from different petroleum based feedstocks (Table 1.1). Monomers, compounds which can be used to react with itself or another similar compound or molecule, form polymers. The feedstocks shown in Table 1.1 are either monomers or used to form monomers. For example, ethylene is a monomer for producing polyethylene, and benzene is used to produce styrene, the monomer for polystyrene.

Polymer resins are formed when a chemical reaction takes place in which the molecules of a relatively simple substance (the monomer) are linked together to form large molecules whose molecular weight is a multiple of the monomer (termed polymerization). In short, polymerization is the bond forming reaction between small molecules with the potential of proceeding indefinitely. Almost all polymerizations are exothermic, and the heat of reaction must be removed in order to maintain the process. Polymerization can occur using a number of methods and can be carried out in batch or continuous operating modes in reactors. The reactor maintains different pressure, temperature and catalyst levels necessary to produce different type polymers. Step-growth polymerization, where two components are used in a conventional chemical reaction to form a polymer which grows in steps, is another polymerization process. The most important example of this is the mixing of ethylene glycol and terephthalic acid to form polyethylene terephthalate (PET), widely used in beverage bottles, textile fibers, transparent films, electrical parts and strappings.

A homopolymer is a polymer resulting from the reaction of one monomer, meaning it consists of a single type of repeating unit. A copolymer is the polymerization of two distinct monomers incorporated into the same polymer chain. For example, ethylene and propylene are monomers used to make copolymer PP; homopolymer PP can also be made from propylene. HDPE, LDPE, PS and PVC can be either homo- or copolymers. PET is a copolymer. Copolymerization is the most general and powerful method of effecting systematic changes in polymer properties, and it is widely used in the production of

Table 1.1 Feedstock Chemicals for the Production of High Volume Plastics [U.S. EPA, 1990a]

Feedstock Chemical	Possible Polymer Product [a]
Acetylene	PVC, PUR
Benzene	PS, PUR, ABS
Butadiene	PUR, ABS
Ethylene	HDPE, LDPE, PVC, PS ABS, PET, PUR, Polyesters
Methane	PET, PUR
Napthalene	PUR
Propylene	PP, PUR, Polyester
Toluene	PUR, Polyester
Xylene	PS, PET, ABS, Polyester, PUR

a. Refer to the glossary (Appendix D) for abbreviation and polymer description.

commercial polymers and in fundamental investigations of structure-property relations [Tirrell, 1990]. Different monomers are used to create the same copolymer resin. Copolymerization can widely vary resin properties such as melting point, glass transition temperature, crystallinity, solubility, elasticity and chemical reactivity. The crystallinity of a resin refers to the amount of ordered three dimensional structure (a property which imparts rigidity to a molecular resin structure). For example, PET is referred to as a crystalline polymer (one whose crystalline properties can be controlled by processing). The glass transition temperature (T_g) refers to the midpoint temperature of a temperature band. When the temperature is above the midpoint, the resin structure is unfrozen and in a rubbery state and will eventually melt with increasing temperature. When the temperature is below the midpoint, the resin is in a hard, glassy condition.

Polymers are characterized according to many parameters. Some of the more basic characterizations are the molecular weight (MW) of the polymer produced, density and the melt point. The density is affected by the MW of the polymer produced, among other things. The melt flow index (MFI) is a simplified measurement for determining grades within a polymer type. It is most widely used in classifying polyethylene (PE) resins. The MFI device involves forcing a polymer melt through a specific size orifice die at a fixed temperature with a specific weight piston. The extrusion device is described by ASTM D1238. The weight expelled in a 10 minute period is known as the MFI, which can typically range form 0.1 to 20. The MFI is inversely proportional to MW, with a MFI=20 being a low MW and 0.1 being a high MW polymer. While the MFI is a simple index, it is only a single point approximation for polymer viscosity, and should only be used to compare polymers in the same family. Therefore values for PE should not be compared with PP or PS. However, it is widely used by polymer suppliers to compare grades.

Table 1.2 shows some of the density, melting points and overall properties of the six primary thermoplastics. Following is an overview of the six primary thermoplastics:

Polyethylenes (PE) The basic structure of the homopolymer PE is the chain -$(CH_2$-$CH_2)_n$-, made from polymerization of the gas ethylene, C_2H_4. In commercial PE, n may vary from 400 to more than 50,000. It is the variation of n and the addition of copolymer substitutes which allow for the variation of melt index, density, and numerous physical properties. There are three basic polyethylene types: Low density polyethylene (LDPE), medium density polyethylene (MDPE) and high density polyethylene (HDPE). LDPE has the variations of linear LDPE (LLDPE), Ultra LDPE (ULDPE) and Very LDPE (VLDPE). HDPE also has the variation of ultra high molecular weight polyethylene (UHMWPE), respectively. This discussion is limited to the two LDPE types (LDPE and LLDPE) and HDPE, as the others are not as broadly used. LDPE used in films is usually a homopolymer. PE as a family has become the largest commercially produced resin in the world. LLDPE is characterized by linear molecules without long-chain branches. In contrast, LDPE contains many long chain branches off the backbone molecules. LLDPE is easier and less energy intensive to produce. HDPE contains a small number of chain branches that are introduced by copolymerization. A low melting point and high chemical stability facilitate the processing of HDPE using injection, extrusion and blow molding.

Polypropylene (PP) The basic structure of homopolymer PP is linked propylene monomer. Copolymerization with ethylene also occurs frequently. Hundreds of grades of PP are sold in the U.S. Its density is among the lowest of all plastics.

Polystyrene (PS) PS is produced by the styrene monomer. Styrene is very reactive and readily undergoes homo- and copolymerization. General purpose PS is often called crystal PS because of the clarity of its appearance. Its commercial success is due mainly to transparency, lack of color and thermal stability. There are three commercial grades of general-purpose PS: easy flow, medium flow and high heat.

Table 1.2 Commodity Thermoplastic Characteristics

Resin Type [a]	Density (g/cm^3)	Softening or Melting Range (°C)	Properties/Applications
LDPE	0.910-0.925	102-112	Largest volume resin for packaging; moisture proof, film transparent
LLDPE	0.918-0.942	102-112	Use generally grouped with LDPE
HDPE	0.940-0.960 0.950 (colored bottles) 0.960 (clear bottles)	125-135	Tough, flexible and translucent material, used primarily in packaging; product examples include milk and detergent bottles, heavy-duty films, wire and cable insulation
PP	0.90	160-165	Stiff, heat and chemical resistant, used in furniture and furnishings, packaging and others; product examples include drink straws, fish nets, butter tubs, auto fenders
PS	1.04-1.10	70-115	Brittle, clear, rigid, easy to process, used in packaging and consumer products; product examples include foamed take-out containers, insulation board, cassette and compact disc cases
PET	1.30-1.40	255-265	Tough, shatter and wear resistant, used primarily in packaging and consumer products; product examples include soft drink bottles, photographic and x-ray film, magnetic recording tape, shipment strapping
PVC	1.30-1.35	150-200	Hard, brittle and difficult to process, but processed easily using additives; a wide variety of properties and manufacture techniques are possible using differing copolymers and additives; product examples include sheet bottles, house siding, cable insulation

a. Refer to the glossary (Appendix D) for abbreviation and polymer description.

Polyethylene terephthalate (PET)	PET is a thermoplastic polyester made by condensing ethylene glycol and terephthalic acid. PET is stable in a wide range of chemicals and possesses good mechanical, electrical, and thermal properties. It has one of the highest densities of the six primary thermoplastics.
Polyvinyl chloride (PVC)	Vinyl chloride is the monomer used to produce homopolymer PVC. It can also be copolymerized with many monomers to produce polymers with a wide variety of properties. PVC is popular because of its versatility. PVC can be injection, extrusion or blow molded.

Additives

Plastic additives are categorized as being either a modifier or protective additive. They are used to alter physical appearance, to alter physical properties of plastics or to retard (or stop) an undesired chemical reaction. Common modifying additives are reinforcing fillers, non-reinforcing fillers, plasticizers, colorants, blowing agents (for foaming) and impact modifiers (for toughness). Protective additives protect the polymer or stabilize it. Examples include UV stabilizers, heat stabilizers, inhibitors to retard degradation and antioxidants. Seventy-five percent of the additives produced are either fillers or plasticizers.

Compatibilizers are a family of additives that have a direct affect on the recycling of plastic. Compatibilizing agents can allow for a bonding between two otherwise unadhering polymers when blended together. Deficiencies in the properties of a resulting polymer mix can be overcome by compatibilizer addition. Polymers may be grafted with a compatibilizer to increase recyclability. A grafted polymer refers to a polymer comprising molecules in which the main backbone chain of atoms has randomly attached side chains containing atoms or groups different from those in the main chain. The main chain may be a copolymer or or may be derived from a single monomer. Examples of compatibilizers include use of methyl methacrylate (MMA)-grafted PE or chlorinated polyethylene (CPE) for PE/PVC blending, styrene grafted PE copolymers for PE/PS blends and styrene grafted PVC for PVC/PS blends [McMurrier, 1990]. There are also compatibilizers for PS/LDPE blends. The use of compatibilizers then allows for higher end use and larger markets than may be expected without them. The general compatibility of differing polymers is shown in Table 1.3.

Product Manufacture

There are eight basic types of polymer molding: extrusion, pultrusion, blow molding, thermoforming, injection molding and compression molding, transfer molding and rotational molding. Extrusion, blow molding and injection molding are the primary manufacture methods for non-durable goods and waste plastic found in MSW.

Table 1.3 Compatibility of Polymers [McMurrier, 1990] a, b

Polymer Type	LDPE	LLDPE	ULDPE/VLDPE	Ethylene Copolymers	HDPE	PP	EPM/EPDM	PS (general purpose, high impact)	SAN	ABS	PVC	NYLON	PC	ACRYLIC	PBT	PET
LLDPE	1															
ULDPE/VLDPE	1	1														
Ethylene Copolymers	1	1	1													
HDPE	1	1	1	1												
PP	4	2	(1)	2	4											
EPM/EPDM	4	4	(1)	3	4	1										
PS (general purpose, high impact)	4	4	4	4	4	4	4									
SAN	4	4	4	4	4	4	4	4								
ABS	4	4	4	4	4	4	4	4	1							
PVC	4	4	4	(2)	4	4	4	4	2	3						
NYLON	4	4	4	(1)	4	4	(1)	4	4	4	4					
PC	4	4	4	4	4	4	4	4	2	2	4	4				
ACRYLIC	4	4	4	(3)	4	4	4	4	4	4	4	4	4			
PBT	4	4	4	(2)	4	4	4	4	4	4	4	4	1	4		
PET	4	4	4	(3)	4	4	4	4	4	4	4	3	1	4	1	
SBS	4	4	4	4	4	4	1	3	2	3	3	4	4	4	4	4

a. Refer to the glossary in Appendix D for polymer description.
b. Compatibility designations: 1=excellent, 2=good, 3=fair, 4=without compatibility. Designations (1), (2), (3)=compatibility level, depending on composition.

Extrusion Molding

Extrusion forces a plastic or molten material through a shaping die on a continuous basis (Figure 1.1). The feedstock may enter the extruder device in a molten state, but generally consists of solid material that is subject to extruder melting, mixing and pressurization. The feed may be powder, pellets, flake or reground material. Most plastic extruders incorporate a single screw rotating in a horizontal cylindrical barrel with an empty port mounted over one end and a shaping die located at the discharge end. A typical simple screw design for a solid fed single screw extruder must convey the plastic entering the screw into a heated compacted environment where the shear force developed by the

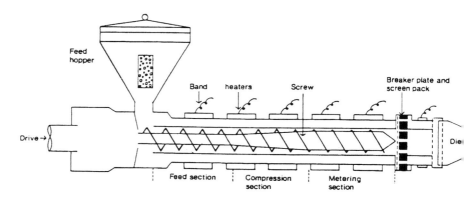

Figure 1.1 A Single Screw Extruder [Morton-Jones, 1989]. The machine consists mainly of an Archimedian screw fitting closely in a cylindrical barrel with just enough clearance to allow its rotation. Solid polymer is fed in at one end and the profiled molten extrudate emerges from the other. Inside, the polymer melts and homogenizes. Twin screw extruders are used where superior mixing or conveying is important.

rotating screw melts the plastic and mixes it to a reasonably uniform temperature while pressurizing the melt and pumping it uniformly through the die.

Both blow molding and injection molding are downstream extensions of the extruder and are discussed separately. Profile extrusion is the direct manufacture of a product from the extruder die. The products are a continuous length with a constant cross section. Examples of profile extrusions include plastic pipe, garden hose and PVC window framing. Profile extrusion is the method now being used to create plastic lumber from mixed plastics. Extruders can also be arranged as coextruders where different layers of polymers are forced through the same die. This allows for multiple layers of polymers to form sheets, films or foam core products. It also allows for sandwiching of barriers (such as a gas permeation barrier) into a product.

Blow Molding

Blow molding produces hollow objects and is performed in three ways: extrusion blow molding, injection blow molding and stretch blow molding. In extrusion blow molding, the melted resin is extruded as a tube into the air (Figure 1.2a, 1.2b). This tube, called a parison, is captured by two halves of the bottle blow mold. A blow pin is then inserted into the mold and air pressure blows the melt against the cavity and cools it. The mold opens, the bottle is ejected, and the bottle trimmed. Extrusion blow molding is generally used for bottles greater than 0.25 liters. Coextrusion blow molding is also being used on a mass production basis to encapsulate scrap/recycle plastic into products. Coextrusion entails multiple extruders feeding resin into separate chambers within the die head used to manufacture a product (Figure 1.3).

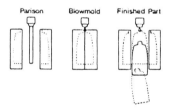

Figure 1.2(a) Extrusion Blow Molding of a Bottle [Miller, 1983]. The parison (plastic hollow tube) is extruded from the die head, the molds close, the part is blow molded, the molds open, and the part is ejected.

104   Mixed Plastics Recycling Technology

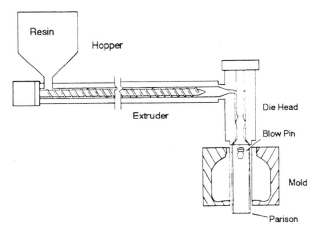

Figure 1.2(b) Extrusion Blow Molding of a Bottle [Miller, 1983]. Hopper, reciprocating extruder screw, die head, extruder parison, and molds shown in open position. The extruder rotates and reciprocates continuously, providing continuous mixing of resin with intermittent extrusion of parisons.

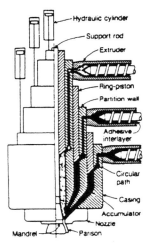

Figure 1.3 Coextrusion Three Layer Blow Molding Die Head [Miller, 1983]. Three independent extruders shown force melt into the die head. Coextrusion is used to install barriers to prevent permeation across bottle walls, add color onto a base polymer, and sandwich recycled resin in between virgin resin.

Introduction   105

In injection blow molding, the melt is injected into a parison cavity around a core rod (Figure 1.4). The test-tube shaped parison, while still hot, is transferred on the core rod to the bottle blow mold cavity where the bottle is blown and cooled. Injection blow molding is generally used for bottles less than 0.5 liter in size. This type of blow molding allows for a scrap free product and for design of intricate shapes such as tamperproof closures. It is impractical for containers with handles.

In stretch blow molding, a heated pre-formed melt is positioned in the blow mold (Figure 1.5). A center rod extends which stretches the preform with axial orientation. Blown air then expands the preform in the mold, forming a bottle with radial orientation. The stretch process takes advantage of the crystallization behavior of the resin and requires the pre-form to be temperature conditioned and then rapidly stretched and cooled. PET soft drink bottles are formed using the stretch method. PET bottles are also formed using extrusion blow molding.

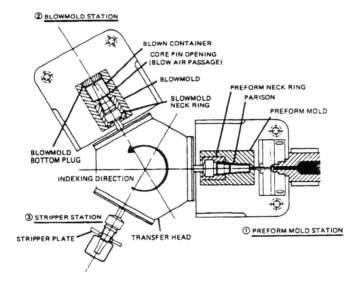

Figure 1.4   Three Station Injection Blow Molding Machine [Miller, 1983]. The parison is injection molded on a core pin (instead of as a tube in free air, as with extrusion blow molding) at the preform mold station (1). The parison and neck finish of the container are formed there. The parison is then transferred on the core pin to the blow mold station (2) where air is introduced through the core pin to blow the parison into the shape of the blow mold. The blow container is then transferred to the stripper station (3) for removal.

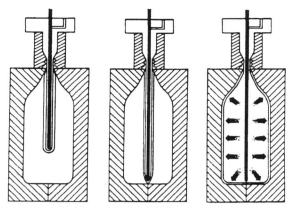

Figure 1.5 Stretch Blow Molding of Bottles [Miller, 1983]. A heated preform melt is positioned in the blow mold. A center rod descends, stretching the preform with axial orientation. Blown air expands the preform into the mold, forming the bottle with radial orientation.

Injection Molding

Injection molding produces solid parts in large volumes at high production rates. It permits close tolerances and the manufacture of very small pieces which are difficult to fabricate in quantity by other methods. Scrap losses are minimal compared to other forming methods because it is one of the only commercial processes where scrap can be reground and remolded at the machine. Melt material is injected into the mold from an extrusion barrel where it is maintained under pressure (Figure 1.6). When the part has sufficiently solidified, the mold opens and the part is ejected.

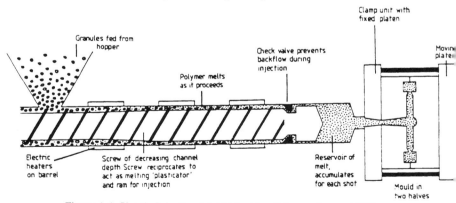

Figure 1.6 Simple Injection Mold Machine [Morton-Jones, 1989]

Film Manufacture

Blown film extrustion and slit die extrusion are the basic methods used to manufacture plastic films. Blown film extrusion produces tubular film and slit die extrusion produces flat film. Tubular film may be cut to produce flat film.

A typical blown film extrusion device is shown in Figure 1.7. The molten polymer from the extruder enters the die and is forced around a mandrel and emerges from a ring shaped die. The extruded film tube is then expanded to a specific diameter by air pressure from the center of the mandrel. As the bubble expands, film thickness decreases. The "frost line" shown in Figure 1.7 is the line where solidification of the extrudate commences.

Slit die extrusion or flat film extrusion as it is also called, is produced by extruding molten polymer through a slit-die and then cooling it on chilled rollers or in a water bath (Figure 1.8). With flat film extrusion, it is important to cool the material within a short distance of the die head so that a clear film is produced and necking (reduction in width) does not occur.

Melt Flow Indexes

Different properties of plastics are preferred for different fabrication methods. The viscosity of the resin melt is typically measured against a standard test referred to as the melt flow index (MFI). The MFI and resin density characteristics of some of the primary applications are shown in Table 1.4. As can be seen, the MFI and density properties vary with fabrication method. To allow for high tolerances and precise definition, injection molding utilizes a resin melt which is runny (MFI=5-400) in comparison to blow mold grade resin (MFI=0.01-0.2) or extrusion grade resin (MFI=0.3-2). Blow mold resin tends to be more taffy like so that it will retain thickness while pressure is applied inside a mold. This viscosity difference limits the type of recycling production methods possible because different melt viscosity resins do not mix homogeneously. If different melt viscosity resins are not separated pror to reuse, the resin properties will not be uniform in production. There is currently no automated method to separate differently produced types of plastics.

108 Mixed Plastics Recycling Technology

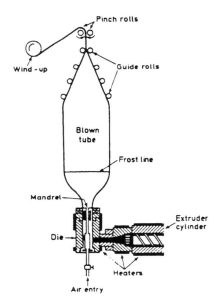

Figure 1.7 Blown Film Extrusion of Tubular Film [Briston, 1989]

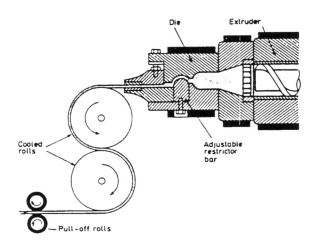

Figure 1.8 Slit Die Extrusion of Flat Film [Briston, 1989]

Table 1.4 Densities and Melt Flow Indexes of Production Polymers

Resin Type [a]	Density (g/cm^3)	Melt Flow Index (g/10 min)
LLDPE film	0.918-0.927	0.1-2.5
LLDPE injection molding	0.915-0.928	5-400
LLDPE pipe and sheet	0.928-0.942	0.8-2.0
HDPE injection molding	0.958-0.968	5-400
HDPE blow molding	0.956-0.964	0.01-0.2
HDPE film	0.946-0.955	0.02-0.3
HDPE pipe and sheet	0.946-0.969	0.3-3.0
PP sheet extrusion	0.902-0.903	0.4-0.8
PP injection mold, general purpose	0.903	4-12
PP injection mold, thin complex parts	0.902	35
PS easy flow injection mold	1.04-1.08	16
PS medium flow injection mold	1.04-1.08	7.5
PS high heat injection mold	1.04-1.08	1.6

a. Refer to the glossary (Appendix D) for abbreviation and polymer description.

2. Manufacture of Plastic Lumber Using Mixed Plastics

The manufacture of flow molded linear profiles, or plastic lumber as it is commonly referred to, has received a great deal of attention as a solution to using mixed plastics because minimal separation of the mixed plastic is necessary to produce this type of product. It is viewed as a method to utilize plastic containers and films en masse which could not otherwise be collected in significant quantity to justify separation. Plastic lumber is also viewed as a method to utilize "tailings," the miscellaneous plastics left after a recycle stream has been "mined" of higher value HDPE and PET bottles. Tailings may also typically be the plastics collected by recycling that were not asked for. Although, from a polymer science point of view, such a diverse combination of plastics is not considered to be readily capable of "blending" into a compatible product, the mixture can easily be processed into large cross-section items that have significant strength and utility [Nosker et al., 1990].

Although the manufacture of plastic lumber from mixed plastics without separation (theoretically) has large potential as a solution to mixed plastics in general, there are associated problems. Depending on market prices and proximity to the manufacturer, it may be necessary to pay a manufacturer to take the waste plastic (there are no mixed plastic lumber producers in Illinois). The cost of shipping can have a large impact on the recycling operation economics. As a result, there are plastic lumber startup companies in progress in Illinois, and Amoco Chemical Company has provided funding to the Center for Neighborhood Technology in Chicago to study and solicit proposals for a mixed plastic recycling plant in the Chicago area [CNT, 1990]. It may also be necessary to separate plastics to obtain a desired color or appearance of the finished lumber product, or to attain a product with reasonable quality standards. While dark browns, blacks and grays are possible with mixed plastic bales, lighter colors such as blue, yellow and light gray are not possible without using separated clear and white HDPE/LDPE. A large proportion of LDPE, both granulated and molded, produces articles which are very elastic. Similarly, a large proportion of PP will produce articles which are brittle. Consequently, blending of granulated material by plastic type may be important depending on the product to be manufactured. If separated with enough quality control, the separated plastics will bring a better price through some other market. The manufacturer may require that a municipality collecting mixed plastics buy the product following recycling.

2.1 Plastic Wood Producers

There are approximately 9 companies currently producing lumber from mixed plastic collection programs, and more companies are reported to be starting up. Three are in the Midwest region: American Plastics Recycling Group (Ionia, MI), Hammer's Plastic Recycling (Iowa Falls, IA) and The Plastic Lumber Company (Akron, OH). The primary manufacturers, including those which take industrial scrap are listed in Table 2.1. Plastic lumber producers which take only HDPE are not listed. As can be seen, some plastic lumber manufacturers have certain conditions or require a certain amount of cleaning beforehand. It should be noted there are also plastic lumber manufacturers which produce only from higher grade recycle stock of natural and colored HDPE bottles which are not shown in Table 2.1.

2.2 Plastic Wood Production

A brief summary of some of the plastic lumber processing methods and the manufacturers in the field follows.

Some of the companies listed in Table 2.1 have developed a proprietary machine for production, while others have purchased the commercially available ET-1 machine manufactured by Advanced Recycling Technologies, Ltd of Belgium. All fabrication processes by the producers shown in Table 2.1 are generally similar and capable of producing thick-wall moldings (park bench pieces and pallets, for example) and profile extrusion pieces (long, straight, thick, heavy, lumber like products). The ET-1 machine accepts any type post-consumer plastic (see section 2.3). Rigid plastic containers must be ground to a 1/4" flake or chip, and film plastics or thin sheet plastics must be densified into small granules to maintain friction in the extruder [Mackzo, 1990]. The regrind and granules are then blended, during which additives and colorants may be added. In addition to post-consumer plastics as a source, packaging scrap, automotive and electrical scrap are cited as potential sources of feedstock.

The ET-1 machine consists of an extruder, molding unit, part extractor and controls. The extruder consists of a short, adiabatic screw rotating at high RPM for melting. The machine is designed to develop thorough mixing of material and to prevent degradation of plastics which are sensitive to heat by providing for a short melt history. The extruder operates at melt temperatures of 360 to 400°F. The temperature is regulated by external cooling, RPM adjustment, or barrel tolerance. Higher melt plastics such as PET and PC, and contaminants such as aluminum and copper become a filler in the melted resins. The molding unit consists of linear molds mounted on a turret that rotate through a

Table 2.1 Mixed Plastic Lumber Manufacturers

Company	Address	Plastic Accepted	Comments
American Plastics Recycling Group	P.O. Box 68 Ionia, MI 48846 (616) 527-6677	65% post-consumer 35% industrial scrap	All rigid plastic containers; accepts films separated out and separated clear and colored HDPE; will accept mixed bales if HDPE bales go to them also, but not alone
Hammer's Plastic Recycling Corp.	RR 3, Box 182 Iowa Falls, IA 50126 (515) 648-5073	60% post-commercial 20% industrial scrap 20% post-consumer	All rigid plastic containers; reluctant to accept film unless separated from rest; community must take product fabricated back; mixed plastic may have to be separated for product
Innovative Plastic Products	P.O. Box 898 Greensboro, GA 30642 (404) 453-7552	Industrial scrap	Accepts only packaging and film scrap from industry; full start-up in early 1991
National Waste Technologies	67 Wall St. New York, NY 10005 (212) 323-8045	99% post-consumer	
The Plastic Lumber Co.	209 S. Main Akron, OH 44308 (216) 762-8988	post-consumer industrial scrap post-commercial	HDPE rigid plastics and LDPE films separated, washed and ground
Plastic Recyclers	58 Brook St. Bayshore, NY 11706 (516) 666-8700	50% post-consumer 50% industrial scrap	
Superwood Ontario	2430 Lucknow, Mississauga, ONT. CA Unit #1 5S1V3 (416) 672-3008	90+% post-consumer 10% industrial scrap	Any plastic accepted; containers and films should be washed out and rinsed; no separation is necessary; future suppliers may have to buy product back
Superwood Alabama	P.O. Box 2399 Selma, AL 36702-2399 (205) 874-3781	40% post-consumer 60% industrial scrap	Accepts separated clear and colored HDPE, and PP (must be separated); separated out films also accepted
Polymerix/ Trimax Plastic Lumber	#4 Frasetto Way Lincoln Park, NJ 07035 (516) 471-7777	85% post-consumer 15% post-commercial	Accepts mixed plastic and/or HDPE, depending on production level and influx from existing accounts

water cooling tank. The extraction unit ejects air from the open end of the mold as it cools and then ejects the part onto an open shelf for removal. The mechanical properties of the resulting lumber (from tailings) are shown in Table 2.2. The mechanical properties of the product from unwashed mixtures are reported to be remarkably consistent even though variations in material collected affect feedstock composition [Renfree et al, 1989]. Capacity of the machines are 300 to 500 pounds/hour.

Table 2.2 Mechanical Properties of Plastic Lumber Profiles [Phillips, et al., 1989, Carrier, 1989]

Testing Group	Composition	Specific Gravity	Compressive Modulus (psi)	Yield Stress (psi) (@2% offset)	Compressive Strength (psi) (@10% strain)
CPRR [a]	100% tailings	0.931	89,600	2,707	3,167
CPRR	PE/heavily plasticized PVC and wire fragments/cable scrap	1.12	35,000	675	1,500
CPRR	50% milk bottles 50% densified PS	0.806	164,000	4,100	4,120 (@ 4%)
Innovative Plastic Prod.[b]	Mostly PE	-	66,000	-	5,800

a. Rutgers University Center for Plastics Recycling Research (CPRR).
b. Innovative Plastic Products additionally reports the following properties: Izod impact strength=105 ft-lbs./inch; tensile strength=2,350 psi; flexural strength=1,950 psi; coefficient of thermal expansion=6.94×10^{-5} in./in./°F.

Another of the plastic lumber manufacturers, Superwood Holdings PLC, licenses its patented process. There is currently one Superwood plant in Canada and another being started in Alabama. The process machinery was developed in Holland and is known as the Klobbie Process. Although HDPE, LDPE and PP are the primary raw materials for plastic lumber production, PET and ABS are allowed but controlled. Due to the fact that PVC will degrade and produce a poor product if present in a mix as a high proportion, it may only be present in small amounts without special additives being added. If a separate source of PVC is available, it can be used at a lower temperature by itself to produce a quality lumber product. Examples of plastic sources for the Superwood process are manufacturers of plastic articles (films, bags, tableware, toys, trays, various domestic articles), scrap from beverage companies, milk suppliers, packagers, below standard products from production facilities and substandard resin pellets from plastic processors. Once collected and separated (for higher added value products), the plastics are blended together in horizontal mixers for a homogeneous mix, conveyed over a magnetic separator and sent to the extruder. A force feed drive may be used to feed the extruder with low density material such as film.

The Klobbie process for Superwood is similar to other mixed plastic lumber methods. Its primary component is the extruder (a large steel screw in a barrel driven at an RPM sufficient to cause friction to melt the resin mix). The plasticized material is forced out through an orifice into a steel mold. The addition of blowing agents and/or fillers vary the characteristics in the end product. The process then changes from extrusion to flow molding. Ten molds of similar or different cross-section but of the same length are mounted horizontally on a carousel in a tank of water. The molds rotate about a horizontal axis and at top dead center each mold is filled by the extruder. The other molds are cooled under water while the top mold is filled and, after being cooled, ejected pneumatically. The water is in a closed system and cooled by a chiller. Additional description of the process may be found in Mulligan, 1989 and Curlee, 1986.

The two Superwood plants in Alabama and Canada are equipped with one machine each, capable of handling 400 pounds/hour. Required capital investment is listed at $2 million per plant [PRU, 1990]. Labor requirements for the Superwood machines are listed as an operator with one to two assistants for the actual machine, two to three personnel for sorting and granulating per machine, and additional help for fabrication of finished products depending on the type of business and the product made [Mulligan, 1989].

The patented Hammer's Plastic Recycling process from Iowa is not sold, but rather operated as a joint venture. The process accepts post-consumer plastics and industrial/commercial scrap discards. A typical blended mixture of plastic prepared for final processing is 65% LDPE, 20% HDPE, 5% PP, 5% PET and 5% miscellaneous. The extrusion machine consists of a fill sensing device, heated extrusion nozzle, screen apparatus and an automatic sprue cutting device [Hammer, 1989]. Closed molds are filled under pressure ranging from 100 psi to 600 psi. Machine throughput is 800 - 1,000 pounds/hour.

Polymerix, a company which produces plastic lumber directly intended to compete with chemically treated outdoor grade lumber, has established a wash/clean process to obtain higher quality plastic with less foreign matter for input to fabrication. The result of material cleaning and partial separation prior to use is a product with mechanical properties similar to wood and better than unprocessed plastic lumber (discussed in section 2.5). A wastewater treatment system for the wash process has been developed for a Polymerix lumber plant. It includes a fixed film upflow biological reactor and UV sterilization.

2.3 General Guidelines for Plastic Lumber Manufacturing

The Advance Recycling Technology ET-1 machine handles a wide variety of thermoplastics, although there are limitations due to the process and specific resin properties, as with the Superwood machine. The following guidelines for the most popular plastics have been submitted by the U.S. supplier of the ET-1 [Mackzo, 1990]:

- *LDPE or LLDPE* A good material for use in the process. However, LDPE is relatively soft and products containing too much of it may be insufficiently rigid for some applications, particularly in thin sections. It should be mixed with stiffer materials such as HDPE or PP.

- *HDPE* A good material for use in the process. HDPE is stiff and its mixtures with LDPE give a range of stiffness that cover most product requirements. Much of the HDPE on the market is copolymer material, but this is of no consequence to the recycler because for recycling purposes its performance is very similar to to that of a homopolymer.

- *PP* A good material for use in the process. It is relatively stiff and its mixtures with LDPE cover most of the range of stiffness requirements. However, the use of more than 30% by weight homopolymer PP is not advised because it is brittle at low temperatures and difficult to nail.

- *PVC* When finely ground and well homogenized PVC can be recycled on the ET-1. It can be mixed with other thermoplastics up to 50% by weight. Post-consumer plastic typically contains 5% PVC or less.

- *PS* Up to 40% by weight of this material can be mixed in. Impact grades add toughness to the mix. Non-impact PS (crystal) tends to cause surface finish problems. Expanded PS (EPS) should be avoided as a foam because of its low bulk density. Testing shows considerable strength improvements at 10 to 40% levels of densified EPS.

- *ABS* A good material for use in the process. The ABS family of resins combines rubbery and plastic properties and is extremely tough. ABS plastics are not broadly available.

- *Nylon* A wide variety are currently on the market. The most common, nylon 6 and 6/6 can be an additive at up to 10% by weight because they impart stiffness to an otherwise soft compound. Textile nylon scrap is usually nylon 6 or 6/6. Nylon 6 castings are suitable. Nylons 11 and 12 are even more suitable, but generally not available.

- *PET* Although its 500°F melt temperature is above the normal range of the ET-1, up to 15% can be mixed in if finely ground and carefully blended. PET beverage bottles, with HDPE base cup, labels and aluminum caps have been run at 100%, but the product is brittle due to crystallization caused by slow cooling of thick sections and degradation of the polymer caused by moisture content.

2.4 Products From Mixed Plastic Lumber

Plastic lumber, as may be expected, has limited applications and use. It costs more than similar outdoor grade products/lumber and is therefore harder to sell, but offers superior resistance to degradation and weathering outdoors. There are transportation, industrial, marine and agricultural applications for plastic lumber products. Such products are generally not yet sold at a consumer level. A number of markets for plastic lumber products are tied to the activities of state and municipal agencies. The following marketplaces are estimated to have the most potential:

- Marine - docks, pilings, seawalls
- Industry - pallets sold directly to end users
- Local authority - transportation/road related markets and barriers, recreational area furniture and pilings
- Agriculture - stakes, electric fenceposts and confinements sold directly by cooperatives
- Fencing contractors
- Public utilities - water, sewage and telephone markers, underground cable covers
- Builder suppliers and garden nurseries - 'do-it-yourself' fencing and stakes

2.5 Enhancement of Plastic Wood Properties

The Center for Plastics Recycling Research at Rutgers has examined uses for plastics tailings (leftovers from recycling after the primary constituents of clear HDPE bottles and PET bottles have been recovered). A significantly greater market exists for the above mentioned separated HDPE and PET beverage bottles than in mixed form, and therefore their recovery may make economic sense. The researchers in New Jersey also have found that 20% of the plastic collected can be tailings material.

It was felt that PS could enhance the properties of lumber product made from tailings which were characterized as largely polyethylenes (HDPE or LDPE). The addition of densified reground PS to the mixture significantly improved the mechanical properties of the resulting product. Mechanical property test results of various additions of PS are shown in Table 2.3. It shows that addition of 10% by weight PS increased modulus of elasticity 60%, yield strength by 15% and compressive strength by 2%. A straight line fit of the data shows that compressive and yield stress increase approximately 15% and 20%, respectively, for each 10% increase in PS up to 50% by weight. Above 35% PS, the modulus of elasticity decreased slightly and appeared to level off around 220,000 psi. A possible explanation for the significant increase in properties of the base material is that PS, which is a glossy polymer at room temperature, reinforces the largely polyolefin matrix in a manner similar to that of fillers used in composite materials even though PS and polyolefins are generally considered to be incompatible [Nosker et al., 1990].

An example of static bending deflection loading of various plastic wood products manufactured by Polymerix is shown in Figure 2.1. This company's products are upgraded from standard extruded plastic lumber with glass fiber reinforcing and a foamed core. It provides deflection load properties similar to the southern yellow pine (Figure 2.1) which is the company's primary wood competitor. The tests were run on a standard 2x4 piece of lumber. Unreinforced, unfoamed plastic lumber has deflection load properties that are less than that of southern yellow pine (Figure 2.1).

Table 2.3 Changes in Mechanical Properties of Plastic Tailings with Addition of Polystyrene [Nosker et al., 1990]

Composition (%)		Compressive Modulus (psi)	Yield Stress (psi)	Compressive Strength (psi)
0 PS	100 tailings	90,000	2,700	3,170
10 PS	90 tailings	144,370	3,100	3,,220
20 PS	80 tailings	163,390	3,860	3,860
30 PS	70 tailings	197,600	4,350	4,350
35 PS	65 tailings	239,000	4,950	4,950
40 PS	60 tailings	222,300	4,750	4,750
50 PS	50 tailings	220,000	5,320	5,320
Standard Virgin Resins: [a]				
HDPE		90,000-150,000	-	2,500
LDPE		20,000-27,000	-	900-2,500
PP		100,000-170,000	-	4,000
PS		450,000	-	6,000-7,300

a. From Perry's Chemical Engineers Handbook, Sixth Ed. Modulus in tension shown.

118 Mixed Plastics Recycling Technology

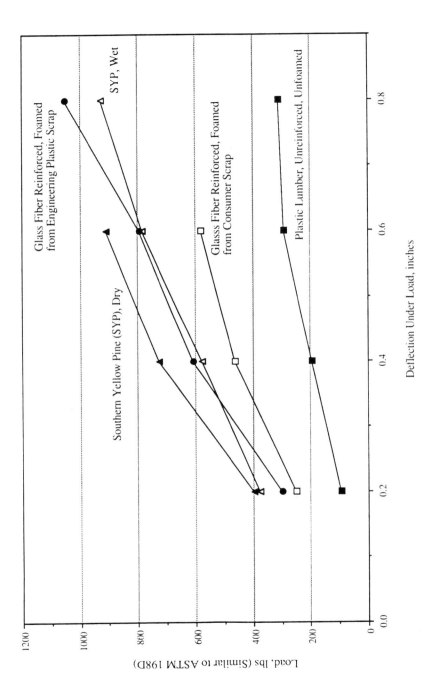

Figure 2.1 Static Bending Test Results of 2x4 Lumber Pieces [Mack, 1990]

2.6 Wood Fiber - Resin Composite Lumber

Wood fiber - recycled plastic composite lumber is a relatively new development and is similar to PS addition to plastic lumber because the potential exists for improved mechanical properties and dimensional stabilities. Wood fiber addition can serve as an excellent reinforcement for plastics, and wood fibers are abundant, lightweight, nonabrasive, nonhazardous and inexpensive. An effort is being made to examine the incorporation of treated and untreated aspen fibers into recycled HDPE milk bottles [Yam et al., 1990]. The work has shown tensile strength and Izod impact strength below that of HDPE alone when aspen fiber is added. Tensile modulus and flexural modulus are increased significantly above HDPE alone with the addition of wood fiber. Dispersion of the fiber in the resin has been reported as a dominating factor in the process thus far.

Another effort is being made to examine the recycle of wood fiber-PS composites into the same material and testing the resultant product under extreme conditions (e.g., exposure to boiling water, at room temperature, 105°C and -20°C). The composite material was reground to a number 20 mesh size and remolded three times in the experiment. Compared with the original extruded composite, the mechanical properties and dimensional stabilities of the recycled material did not change significantly even after exposure under extreme conditions. Detailed data on the study are contained in Maldas and Kokta [1990].

2.7 Future of Mixed Plastic Lumber

Whether mixed plastic lumber survives as part of the long term solution to post consumer waste plastic is unclear. There are currently a number of hurdles with such a product. It is uneconomical to ship relatively low density low-value waste plastic more than 100-200 miles for production. There are clearly technical advantages of plastic wood in certain applications, and therefore its manufacturing cost must be compared to similar outdoor grade wood products. An effective infrastructure for collection, transport, processing and marketing of plastic lumber product is necessary in the area where waste plastic is generated in order for plastic lumber to be a long term viable alternative. For Illinois, the best location for such a facility would be the metropolitan Chicago area. A metropolitan area can provide consistent volumes of feed material to support an operation 24 hours a day, 7 days per week and allow for the removal of high grade, high priced waste plastics from the waste stream while processing the "tailings" and providing markets for products.

3. Emerging Methods for Processing and Separation of Plastics

Because the manual sorting of heterogeneous plastic mixtures is the current level of technology in all but a few applications of plastics recycling, methods to effectively separate them on a more automated basis is receiving attention. Methods for automated sorting rely on such responses as specific gravity changes, x-ray diffraction, optical recognition and dissolution in solvents. No one of the above methods can completely sort any type of plastic mixture.

The sorting methods being worked on can be classified into macro, micro or molecular scale sorting. Macro sorting involves separating plastic based on an entire product such as using optical sensing to separate whole bottles by color. Manual separation is also a macro sorting method. Micro sorting involves the initial processing to a uniform criteria, such as size, with subsequent separation. An example of micro sorting is the commercially viable PET bottle/HDPE base cup separation method where bottles are ground up and sent through a hydrocyclone, the PET and aluminum cap grind sinks while the HDPE floats. Molecular separation involves processing plastic by dissolving the plastic and then separating plastics based on temperature.

3.1 Optical Color Sorting of Glass and PET Containers

This method currently exists as a laboratory prototype with a larger prototype being designed and fabricated for a local recycling center. The clear/color sorting system (CCSS), being developed at the University of Illinois Urbana-Champaign (UIUC), is for the clear/color separation of glass and PET plastic beverage bottles using optical sensors coupled with a control circuit to activate diversion of clear glass or PET from colored glass and plastic.

While processing glass at the Champaign, Illinois Recycling Center, the CCSS will be a separation aid in that the automated sorter will divert a majority of clear glass containers from the glass recycle stream while leaving green and amber glass containers. A manual sorter will then have to remove brown bottles from the green glass stream. A simplified diagram of the container separation system is shown in Figures 3.1(a) and 3.1(b). The laboratory prototype consisted of a manually fed trough followed by an alignment device which aligned bottles so they would not jam the conveyor and would pass over the optical sensing device. Directly below the alignment section was the optical sensing section and then the diversion section, which consisted of a plywood diverting gate moved by an electric solenoid with a one inch actuation stroke [Newell and Lewis, 1990].

Emerging Methods for Processing and Separation of Plastics   121

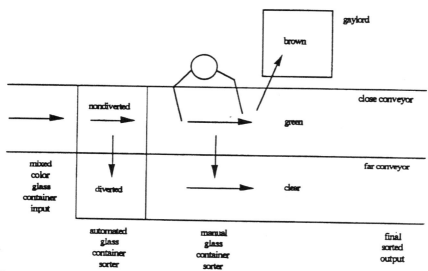

Figure 3.1(a) Clear/Color Sorting System Picking Station Arrangement [Newell and Lewis, 1990]

122   Mixed Plastics Recycling Technology

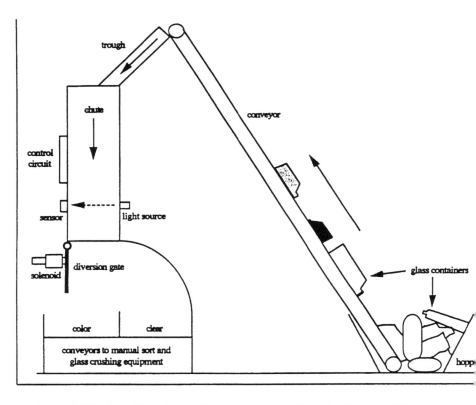

Figure 3.1(b) Clear/Color Sorting System Schematic [Newell and Lewis, 1990]

The researchers reported difficulties with alignment of PET bottles due to their light weight. There was also some problem with glass bottles such as long neck beer bottles, small baby food bottles and large apple cider jugs either not being fully diverted far enough away (specifically large bottles) or not properly passing through the feed/alignment chute. Development of a better alignment system has been indicated for the larger scale prototype.

Test sorting runs on glass bottles using the laboratory prototype ranged from 2,118 to 7,660 bottles per hour. An overall ratio of 60% clear to 40% colored containers was used because it represents that which is found in the waste stream. Results of different sorting speeds and color arrangements tested with glass bottles are shown in Table 3.1. There were no reported problems with the sensing or diverting system, but generally unsuccessful runs were reported to occur due to the trough feed chute (bottles ganging up at the chute entrance). The overall results show that 87% of the containers were separated correctly, 98% of the clear stream had clear glass containers and 76% of the colored stream had colored glass containers. The mass rate of glass can be estimated using 0.5 lbs/bottle for all runs except drop no. 1 (0.481 lbs./bottle) and drop nos. 11-13 (0.469 lbs/bottle).

The test results for PET bottle processing are shown in Table 3.2. The optical sensing mechanism for glass was reported to be compatible for PET bottles without recalibration. Smaller one liter PET bottles jammed at the chute entry, as previously indicated. Ninety-five percent of the processed clear PET bottle stream contained clear bottles while 100% of the color stream contained colored PET bottles when considering bottle runs where alignment was not a problem. An estimate of the mass flow rate can be made using 0.113 lbs/1 liter bottle and 0.163 lbs/2 liter bottle.

A primary feature of the sorting method, in addition to high separation efficiency and simple design, is the low capital and operating cost of the system. It is projected to cost $2,000 to $4,000 to install a unit which processes 1,000 tons/yr glass and 60 tons/yr PET, based on the estimated recycling rates for the local Urbana/Champaign recycling center. The total usage time for this amount is 177 hours annually. It is estimated that a minimum of 136 person-hours would be saved annually by the sorter removing 81% of the clear glass automatically. This does not include time to separate color glass form the clear stream and clear glass from the color stream. It is also estimated that the CCSS would reduce the time required to sort PET bottles by 50%. The cost of the feed/conveying system is the most significant portion of capital expense. A facility with 10 times the glass/PET load (10,000 tons/year glass and 600 tons/year PET), which could make better use of operating time, is estimated to result in a savings of $11,000 to $22,000 per year resulting in a payback of less than one year.

Table 3.1 Results of Glass Container Color Sorting Tests [Newell and Lewis, 1990]

Drop no.	Pattern [a] (colors)	Clear bottles triggering solenoid	Color bottles not triggering solenoid	Total % correct response	Total time elapsed (seconds)	Time / bottle (sec/bottle)	Rate (bottles/ hour)
1	10 brown (B)	0/0	9/10	90	5.42	0.542	6,642
2	10 clear (C)	10/10	0/0	100	6.37	0.637	5,651
3	10C	9/10	0/0	90	16.14	1.614	2,230
4	10 green (G)	0/0	9/10	90	10.30	1.030	3,495
5	8C	5/8	0/0	63	7.02	0.878	4,100
6	10C	8/10	0/0	80	11.94	1.194	3,015
7	5B, 5G	0/0	10/10	100	14.45	1.445	2,491
8	5B, 5C	4/5	5/5	90	9.61	0.961	3,746
9	C,B,G,C,B,G, C,B,G,C	4/4	6/6	100	10.15	0.882 [b]	4,082
10	B,C,G,C,B,C, G,C,B,C	4/5	5/5	90	6.86	0.686	5,248
11	10C	8/10	0/0	80	7.60	0.470 [c]	7,660
12	10C	8/10	0/0	80	8.10	0.810	4,444
13	10C	8/10	0/0	80	17.00	1.700	2,118
14	C,C,B,G,C,C, B,B,G,C	5/5	5/5	100	9.94	0.994	3,622
15	B,B,C,B,C,G, C,C,B,C	5/5	5/5	100	8.98	0.898	4,009
16	C,B,C,G,C,G, C,B,C,B,C,G	3/6	6/6	75	12.71	1.059	3,399
17	C,G,C,G,C,B, C,B,C,B,C,B	4/6	5/6	75	14.68	1.223	2,943
18	C,G,C,B,C,B, C,B,C,G,C,G,C	5/7	6/6	85	15.95	1.227	2,934
19	C,C,C,C,B,B, G,G,C,C,C,C,C	7/9	4/4	85	20.50	1.577	2,283
20	B,C,C,G,C,G, C,B,C,C	5/6	3/4	80	17.10	0.874 [d]	4,118
Totals		102/126 (81%)	78/82 (95%)	87			

Clear containers diverted to the clear holding bin = 98%

Colored containers diverted to the colored holding bin = 76%

a. Number, color and order in which bottles were processed are shown (B=Brown, C=Clear, G=Green)
b. Time adjusted from 1.015 sec/bottle due to feed discrepancy
c. Time adjusted from 0.760 sec/bottle due to feed discrepancy
d. Time adjusted from 1.710 sec/bottle due to feed discrepancy

Table 3.2 Results of PET Beverage Bottle Color Sorting Tests [Newell and Lewis, 1990]

Drop no.	Pattern [a] (colors)	Clear bottles triggering solenoid	Color bottles not triggering solenoid	Total % correct response	Total time elapsed (seconds)	Time / bottle (sec/bottle)	Rate (bottles/ hour)
21	5G (2L) 1G (1L)	0/0	5/6	83%	12.82	2.137	1,685
22	4G (2L) 1G (1L)	0/0	5/5	100%	12.33	0.837 [b]	4,303
23	4C (1L) 3C (2L)	6/7	0/0	86%	12.57	1.796	2,005
24	4C (1L) 3C (2L)	4/7	0/0	57%	7.12	1.017	3,539
25	G,C,G,C,G,C G,G	3/3	4/5	88%	11.46	1.433	2,513
26	G,C,G,G,C,G, C,G	3/3	5/5	100%	8.13	1.016	3,542
27	C,C,G,C,C	1/4	1/1	40%	4.79	0.958	3,758
28	G,G,C,G,G,C, C,G	3/3	5/5	100%	9.64	0.894 [c]	4,026
29	C,G,G,G,C,G, C,G	3/3	5/5	100%	6.11	0.764	4,714
30	G,C,C,G,G,C, G,G	3/3	5/5	100%	5.58	0.698	5,161
Totals							
	1 liter [d]	5/12 (42%)	2/3 (67%)	47%			
	2 liter	21/21 (100%)	33/34 (97%)	98%			
	Combined	26/33 79%	35/37 95%	86%			

Clear containers diverted to the clear holding bin = 95%

Green containers diverted to the green holding bin = 100%

a. Number, color and order in which bottles were processed are shown (C=Clear, G=Green).
d. Time adjusted from 2.466 seconds/bottle due to feed discrepancy
c. Time adjusted from 1.205 seconds/bottle due to feed discrepancy
d. Minor alignment difficulty occurred with 1 liter containers

3.2 Separation of PVC Bottles from Other Plastic Containers

The separation of PVC bottles from other plastic in sorting programs has become important due to the adverse effects of even small amounts of PVC present in other plastic, such as PET. Upon being melted with PET, which is what PVC is most often mistaken to be, hydrochloric acid can form and corrode the metal parts used in plastic extrusion machines.

Work at B.F. Goodrich Co. and the Rutger's CPRR has yielded separation methods for PVC bottles. The CPRR separation method is based on using x-ray fluorescence to detect characteristic backscattering from chlorine atoms in PVC which is higher than that detected by polyolefin plastic without chlorine. Mechanical separation can then occur. The x-ray detection is reported sensitive enough to trigger removal of a PET bottle with a PVC label, and can also detect other chlorine containing polymers such as polyvinylidene chloride (PVDC) which is used in laminates and films [Monks, 1990]. Although the detection of small amounts of chlorine is possible, the chlorine x-ray is weak and does not penetrate paper labels. There is also a rapid decrease in measured x-ray chlorine intensity as a particular sample is moved away from an x-ray source and detector. This could lead to a potential problem for bottles of uneven shape [Summers et al, 1990a]. Figure 3.2, which shows responses for two different type detectors, indicates that the chlorine reading decreases by about half for every 3 mm of air space between the sample and the detector.

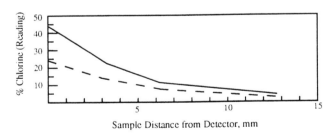

Figure 3.2 Chlorine Detection Strength Versus Bottle Distance from Detector [Summers et al, 1990b] Shown are responses from two detector models

In addition to chlorine intensity being affected by the distance from the detector, the intensity will also vary across the length of a bottle. Figure 3.3 shows the x-ray detector response to a uncrushed bottle and Figure 3.4 shows the response of a crushed bottle, both measured at at 1 cm intervals across the bottle from the face and profile views. Both figures show that a chlorine reading at a minimum of 2 cm intervals will identify the peak reading for the bottle. Based on this limitation it is projected that any bottle should be measured several times at 2 cm intervals. Because the detector can make a reading in 0.005 seconds, a maximum belt speed of 2 cm/0.005 sec, or 400 cm/second is recommended. A 10 cm spacing between bottles would allow approximately 10 bottles/second to be analyzed.

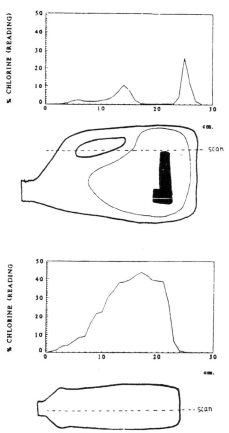

Figure 3.3 X-Ray Fluorescence of PVC Oil Bottle from Front and Profile View [Summers et al, 1990b] The black marking represents bottle labeling

128   Mixed Plastics Recycling Technology

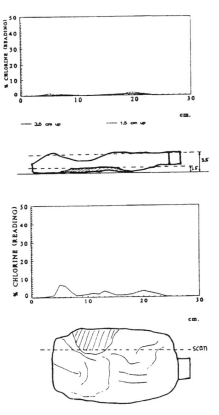

Figure 3.4   X-Ray Fluorescence of Baled PVC Bottle from Front and Profile View [Summers et al, 1990b]

Two other companies, International Food Machinery and National Recovery Technologies (NRT), the latter receiving funding from the U.S. EPA and the Vinyl Institute, also have processes to separate PVC bottles. The exact sensing processes used by these two companies are not available; however, it is believed to be similar to CPRR. A simplified diagram of the NRT sorting system is shown in Figure 3.5. Other companies are marketing a sensing device as well.

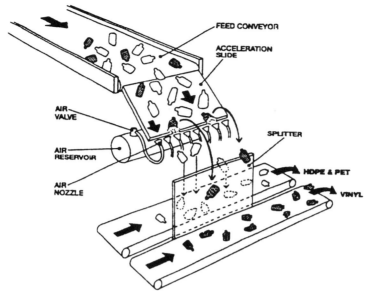

Figure 3.5 National Recovery Technologies Vinyl Separation System

A study of PVC bottles in the Akron, Ohio RPC mixed plastic recycling program showed that a manual separation to achieve a PVC stream will be about 80% accurate in identifying which bottles are PVC. Even with automated x-ray separation, additional cleaning is necessary and a method to perform cleaning and non-specified plastic separation using a 1.35 and 1.30 specific gravity calcium nitrate solution has been proposed as shown in Figure 3.6 [Summers et al, 1990b].

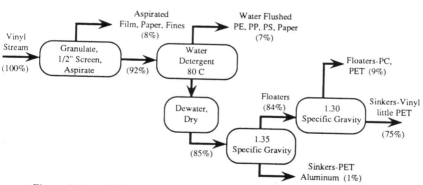

Figure 3.6   Vinyl Stream Purification Process [Summers et al, 1990b]

3.3 Separation of HDPE Base Cups from PET Beverage Bottles

The Center for Plastics Recycling Research (CPRR) at Rutgers University was commissioned to develop a process for the reclamation of soda bottles that could be widely demonstrated and transferred to the public to spur increased recycling in this area. This was done because there are a number of companies in the U.S. which recycle PET soda bottles, but their methods are generally proprietary and not licensed out. The Rutgers Beverage Bottle Reclamation Process (BBRP) is not meant to separate PET soda bottle colors or other type plastics, but rather to provide clean PET flake separated from caps, labels, base cups and adhesives.

Because there are differences in the way PET beverage bottles are produced, it was necessary to determine the composition of soda bottles collected. The primary differences among bottles is the label materials used, cap material used, and whether a HDPE base cup is used. The average composition of PET soft drink bottle constituents are shown in Table 3.3.

Table 3.3 Average Composition of Mixed Soda Bottles Received at Rutgers, NJ During 1988-1989 [Dittman, 1990]

Compound	Average Composition (Weight %)
PET (including green and clear)	73.6
HDPE [a]	20.0
Paper (label) [b]	2.9
EVA [c]	1.3
Aluminum	1.0
PP (labels) [d]	0.8
PP (cap liner)	0.4

a. Actual HDPE quantity is 23% for bottles with base cups and 0% without base cups.
b. Value goes to 0% for bottles with plastic labels.
c. Actual ethylene-vinyl acetate (EVA) quantity, which is a base cup adhesive, is 1% for bottles without base cups and 3% for bottles with base cups. EVA is a copolymer adhesive in the polyethylene
d. Value goes to 0.2% for bottles with paper labels. For those bottles, PP is used for cap liners only.

The process also provides clean polyolefin flakes which are mainly the HDPE base cup and PP caps, and aluminum chips from caps. The process, a schematic of which is shown in

Figure 3.7, is summarized as follows [Rankin, 1990, Dittman, 1990]:

- Collection of waste plastic

- Sortation of waste plastic into uniform types

- Granulation or cutting of the sorted plastics into chips of about 3/8" mesh size. This may be accompanied by air classification to remove loose dirt, paper shreds and other "fines."

- Mixing of the plastic chips with a heated aqueous detergent bath, and agitating the resulting slurry for sufficient time to achieve the desired degree of cleanliness, including label removal, EVA removal and disintegration of any paper present.

- Separation of the wash bath from the plastic chips via a dewatering screening device. Paper fibers and dirt are removed along with the bath, except for that portion which remains clinging to the wet chips. The wash bath is then filtered and reused. Make-up detergent solution is added as required.

- The wet and washed chips are re-slurried in rinse water in an agitated tank, and pumped through a set of hydrocyclones to separate the "light" components (polyethylene base cups and polypropylene labels) from the "heavy" components (PET and aluminum bottle tops). The specific gravities are as follows:

  a) HDPE   0.96
  b) PP   0.90
  c) PET   1.29 to 1.40
  d) Al   2.60

  Thus items a) and b) will float in water, while c) and d) will sink.

- The light and heavy components are individually dewatered to 3-7% H_2O using high-speed rotating fine screens and dried. Thermal separation is used to remove residual water from the two fractions after dewatering. One dryer is used for each stream, with the PE/PP stream dried to 1% H_2O and the PET/Al stream dried to 0.179% H_2O or less. The low moisture content of the latter is necessary for electrostatic removal of the aluminum.

- The light component stream is then ready for sale, or an optional step can be used to remove PP from PE. PE/PP separation can be done by increasing the air velocity in the light stream hot-air drier. PP chips will be carried out by the air. It has been found, however, that most purchasers are satisfied to buy HDPE chips containing 7-8% PP by weight. They can be used as a mixture in injection molding, since their melting point temperatures are not far apart:

  HDPE   130 - 137°C
  PP   168 - 175°C

132 Mixed Plastics Recycling Technology

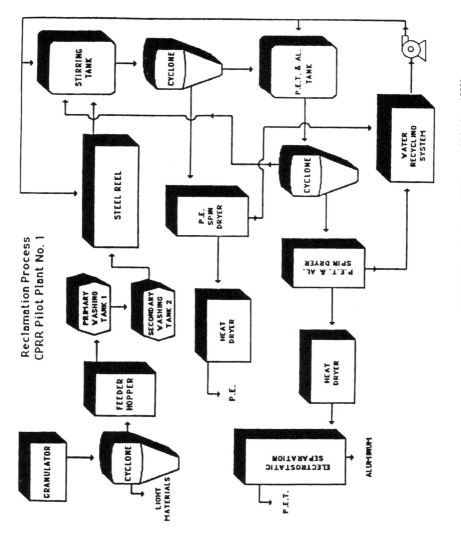

Figure 3.7 Schematic of Rutgers Beverage Bottle Reclamation Process [Morrow and Merriam, 1989]

- The heavy component stream is sent to an electrostatic Carpco separator for removing aluminum. The PET and Al mixture is fed continuously onto a rotating steel roll maintained at high voltage. The electrostatic aluminum removal process takes aluminum from 1.2% (12,000 ppm) to 50-100 ppm residual aluminum in the cleaned PET flake. The aluminum by-product contains approximately 50% Al and 50% PET by weight. An additional smaller separator is necessary to attain a higher purity aluminum material.

Average analysis of the PET flake and HDPE product is shown in Table 3.4. The wastewater from the process should not be much different from a large cafeteria and generally does not require pretreatment. Byproducts are reported to be wet sludge (mainly paper fibers), dirty rinse water (dishwater) and dirty wash solution (concentrated dishwater). Rutgers has installed a 5 million pound per year pilot plant to demonstrate the operation.

Table 3.4 Product Analysis of Rutgers Beverage Bottle Reclamation Process [Dittman, 1990]

Component	Mean PET Product	Mean HDPE Product
PET, including EVA (wt %)	99.756%	0.02 to 0.63%
PET, EVA free basis [a] (wt %)	99.995%	-
HDPE	10.3 ppm	92.4 - 99.4%
PP	1.2 ppm	0.0 - 7.5 ppm
Aluminum	25 - 100 ppm	0.0 - 2.0 ppm
Average moisture content	$\leq 0.25\%$	1%

a. Ethylene-vinyl acetate (EVA) can be removed to as high a degree as desired.

The BBRP license and technology is available for a license issuance fee of $3,000 and a royalty fee equal to 1/4% of the gross sales price of the licensed products that are used, leased or sold by or for the licensee [Rutgers]. After the license agreement royalties reach $25,000, no more royalty fees will be due. The license issuance fee includes a technology transfer manual which provides detailed equipment and process description, process economic estimates, safety and health parameters and quality control requirements and measurements. A detailed cost estimate was performed by CPRR for a 20 million pound per year facility in a leased building which was based on attaining 20¢/lb for PE flake and 34¢/lb for aluminum. Plant operation was based on 24 hours/day, 330 days/year. Investment costs and revenues resulting from varying clean flake PET prices of 31¢/lb, 36¢/lb and 41¢/lb are shown in Table 3.5. It estimates a return on investment of 34%/yr, 50%/yr and 66%/yr for the three PET prices, respectively. Similar analysis for a 10 million pound per year plant shows a return on investment of approximately 4%/yr, 16%/yr and 29%/yr for PET flake prices of 31, 36 and 41¢/lb, respectively [Phillips and Alex, 1990].

Table 3.5 Cost Analysis Summary of Rutgers PET Bottle Processing Plant [Phillips and Alex, 1990]

Description	Value ($M/yr) PET Price (¢/lb)		
	31¢	36¢	41¢
Revenue [a]	5255	5915	6574
Operating cost	-4000	-4000	-4000
Taxes	-464	-708	-952
Profit after tax	791	1207	1622
Total capital investment	2351	2406	2460
Return on investment (%/yr)	34	50	66
Payback period (years)	2.4	1.7	1.4

a. Based on polyethylene product price of 20¢/lb and aluminum price of 34¢/lb.

3.4 Separation Using Selective Dissolution

This method is currently being studied at a laboratory level. Selective dissolution involves the separation of mixed plastics on a molecular scale by dissolving resin mixtures in a solvent. Examining this process as a method to separate mixed plastics was started in the mid 70s. There are two methods being approached in the dissolution process. The first method uses one solvent to dissolve all resin types and the second method uses one solvent to dissolve one particular type of resin, but not others. Both methods have received attention because the plastic stream can be heterogeneous in nature and contaminants such as metals, glass, cellulose and some pigments can be removed. Selective dissolution can allow for microdispersion of polymer combinations, thereby rendering innocuous certain plastic components that may lead to manufacturing difficulties or poor physical properties.

The process using single solvent addition (flowchart shown in Figure 3.8) involves the separation of mixed plastics using two primary steps [Lynch and Nauman, 1989]:

- Selective dissolution - A solvent and a sequence of solvation temperatures are used in a sequential batch mode to selectively extract a single polymer group from the commingled stream. The polymer obtained from the single extraction is isolated using flash devolatilization. The recovered polymer is then pelletized and the condensed solvent is returned to the dissolution reservoir to remove the next group of polymers at a higher temperature. Typical conditions consisted of placing 25 kg of plastic waste (virgin polymers used) into the column with screens at each end. A pump circulated 20 liters of solvent through the heat exchanger and dissolution column.

- Flash devolatilization - Once separated by selective dissolution, flash devolatilization of the solvent from the mixture is used to produce solvent free polymers. This process is an outgrowth of the polymer process of compositional quenching, where two incompatible polymers are dissolved in a common solvent, and then the solvent devolatilized such that phase separation between the two polymers occurs and a microdispersion of one polymer within the other occurs. The microdispersion renders the minor constituents innocuous. Figure 3.9 shows an experimental flash devolatilization apparatus. The polymer concentrations in the solvent are typically 5 to 10% by weight. The pressure in the heat exchanger is sufficient to prevent boiling with pressure maintained by the flash valve at 10 - 40 atm, and a typical temperature upstream of the flash valve of 200 - 300°C. The temperature upstream of the flash valve and the flash chamber pressure (typically 5 - 100 torr) are the two variables which govern the devolatilization step. They determine the polymer concentration after flashing (typically 60 - 95%) and the after flash temperature, 0 - 100°C.

Equal volumes of the six major thermoplastics were used: HDPE, LDPE, PET, PP, PS and PVC. Tetrahydrofuran was selected as the first trial solvent due to the data acquired in previous compositional quenching work [Lynch and Nauman, 1989]. Xylene has also been used. Results achieved are shown in Table 3.6. It shows that a four way split between plastic types can be achieved with good separation efficiencies. A lower separation efficiency is expected with an actual commingled waste plastic stream due to variations in polymer properties among manufacturers. However, it is expected that compositional quenching will overcome the problems. Preliminary economics has indicated a 50 million pound per year plant will process waste plastic for around 15¢/pound [Lynch and Nauman, 1989].

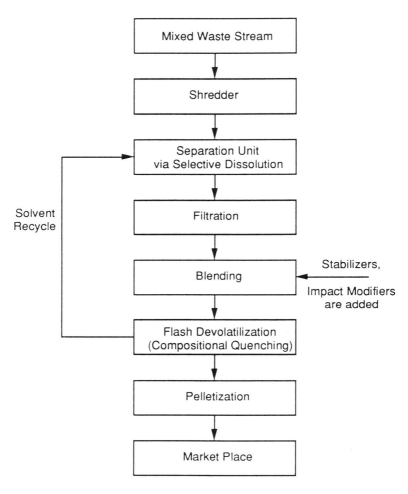

Figure 3.8 Single Solvent Selective Dissolution Process Flow Sheet [Lynch and Nauman, 1989]

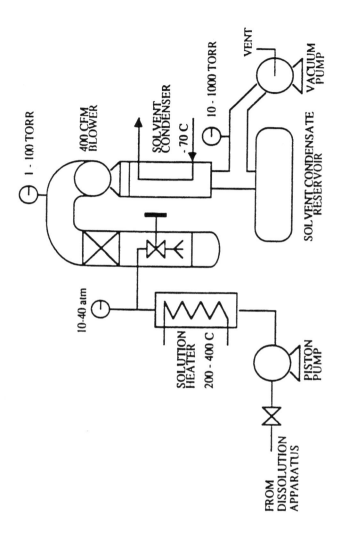

Figure 3.9 Selective Dissolution Single Solvent Flash Devolatilization Miniplant [Lynch and Nauman, 1989]

Table 3.6 Experimental Extraction Efficiencies Using Single Solvent Selective Dissolution [a] [Lynch and Nauman, 1989]

Material	Extraction	Temperature (°C)	Efficiency (%)
PVC	First	25	>99
PS	First	25	>99
LDPE	Second	70	>99
PP	Third	160	>99
HDPE	Third	160	>99
PET	Fourth	190	>99

a. Experiments were conducted using virgin polymers. Tetrahydrofuran was the solvent used in dissolution.

The multiple solvent process involves the use of a solvent compatible with a limited number of polymers. It has advantages over the single solvent process in that lower pressures and temperatures are necessary which results in reduced energy requirements. Because a different solvent is used for each polymer, a potentially higher purity product can be obtained. Work in this area has focused on PET bottle flake purification following mechanical cleaning from other soda bottle constituents of HDPE, PP, paper and aluminum. This purification process would result in a high purity PET polymer at an increased cost.

A flowsheet of the multiple solvent process, shown in Figure 3.10, is described as follows [Vane and Rodriguez, 1990]:

- Stage I - Dried chips from a mechanical, density based, float-sink system (such as the Rutgers BBRP process) are fed in, where they are washed with the process solvent at a temperature sufficient to remove insoluble impurities (up to 130°C), but insufficient to dissolve the intended polymer. For example, in the PET train, this washing will remove any adhesives, PS or PVC which may be present from use in 2 liter bottles or from sortation error.

- Stage II - The intended polymer is dissolved by the process solvent at higher temperatures (around 170°C). Once dissolution is complete, the solution can be purified.

- Stage III - Undissolved materials are removed using sedimentation/flotation and filtration. Dissolved materials, such as dyes, catalysts and inks are removed by adsorbents.

Emerging Methods for Processing and Separation of Plastics   139

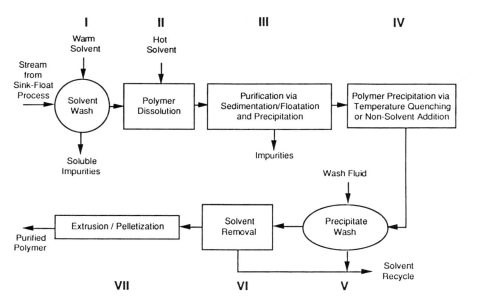

Figure 3.10 Flow Sheet for Multiple Solvent Selective Dissolution Showing Seven Processing Stages [Vane and Rodriguez, 1990] The feed material would be pre-cleaned by a sink-float system such as the the Rutgers Beverage Bottle Reclamation Process

- Stage IV - The polymer is recovered by temperature quenching or by adding the solution to a non-solvent causing immediate precipitation of the polymer.
- Stage V - Rinsing to remove dissolved impurities from the polymer precipitate
- Stages VI and VII - Removal, extrusion and pelletization.

Based on criteria of cost, toxicity, solvent recovery and favorable PET solution behavior, and incompatibility with polyolefins, the solvent utilized for dissolution was n-methyl-2-pyrrolidinone (NMP). Figure 3.11 shows the dissolution rate of bottle PET in NMP. From this figure, the PET dissolution rate is achievable without elevated pressures,

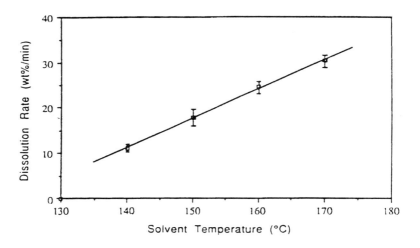

Figure 3.11 Dissolution Rate of Bottle PET in N-Methyl-2-Pyrrolidinone (NMP) as a Function of Temperature [Vane and Rodriguez, 1990]

as the boiling point of NMP is 202°C. At 170°C, a 2 liter bottle will dissolve completely in about 4 minutes. Conversely, HDPE immersed in NMP at 175°C for 30 minutes will hardly be dissolved (0.2% by weight). At 130°C, PET did not dissolve appreciably even after 4 hours in the solvent.

Testing has not reached the level of determining overall process purity. It is estimated that if a 2 liter bottle (assumed to be of a 70% PET, 30% HDPE composition) was immersed in NMP at 175°C for 30 minutes followed by the previous process stages described, the resulting polymer would be 99.91% by weight PET. If the process were preceded by the Rutgers BBRP process, with 90% HDPE removal, PET quality would be 99.991% PET by weight. Although this does not examine other bottle impurities, a 99.99% pure PET resin is obtainable.

3.5 Separation Using Soluble Acrylic Polymers

This work is under development by a private company, Belland AG - Switzerland, which specializes in acrylic polymers. The solubility and subsequent precipitation potential is available by using acrylic plastics. Applications are envisioned in recyclables, protective coatings, disposables and temporary packaging. An important use for such a material would be to make recycling of other products easier without being directly incorporated into the recycling stream. For example, producing telephone books using an easily soluble adhesive binding would allow telephone books to be more readily recycled. A labeling system consisting of alkaline soluble label film stock, adhesive and ink has been developed for this purpose [Wielgolinski, 1990].

The soluble polymers are based on acrylic monomers which contain a definitive percentage of carboxyl groups. The carboxyl additives render the polymers soluble in an alkaline solution such as 2% ammonia solution or a dishwashing detergent. Solubility can then be reversed to precipitate the polymer after lowering the pH of the solution by acidification. The polymer is water repellant and therefore would not be damaged during its normal life. The polymer properties can range from stiff and strong for injection molding, to tough and flexible for extrusion molding, and down to soft and sticky for adhesive applications; over 1,500 different materials have been prepared which differ in acrylic composition, molecular weight and dispersity [Wielgolinski, 1989]. Of these, 25 co- and terpolymers have been chosen as standard products for coatings, adhesives, films and injection and extrusion molding resins.

The application to plastic recycling lies in the separation of a family of different property acrylics by changing the pH of the acrylic solution. A recycling system has been examined for the fast-food industry where all eating utensils, straws, foamed cups, thermoformed cup lids and foamed clam shells are envisioned as being composed of different formulations of alkaline soluble acrylic. This fast food waste would be subject to alkaline solution digestion with mechanical agitation which would dissolve away and separate organic waste from the plastic. The liquor and insoluble solids would then be separated and the solids given two washes for removal of trace solvated plastic. Polymer containing liquors would combined and purified of color factors by adsorption in carbon filters and molecular sieves. The purified acrylic solution would then be fractionated into separate polymers by virtue of their pH sensitivity. The fractions are then dried, re-extruded and recycled into a primary recycling application. The pH sensitivity of each polymer is a function of its percentage carboxyl functionality and other related factors [Wielgolinski, 1989].

The applications to recycling include use as an allowable material which would not hinder recycling (because it would be easily dissolved) and as a recyclable product. A PET beverage bottle which is affixed with such a soluble film plastic label, cup adhesive and ink, would allow for less contaminant in the PET recycle. Contamination of paper label residue and adhesives is a problem with PET bottle recycling. This also allows for recycle of the adhesive and film by precipitation from the alkaline wash.

3.6 Initial Activities in Polyurethane Recycling

There has been little past nationwide activity in the recycle of polyurethanes (PUR) or phenolics even though these resins are among the largest produced. These two resins comprised 9.5% by weight of the resins produced in the U.S. in 1989 (refer to Table 1.2 in Part I). An overview of some recent work is summarized here.

For polyurethanes, the sixth largest resin produced resin in the U.S., the Society of the Plastics Industry has formed a Polyurethane Recycle and Recovery Council (PRRC) to address PUR re-use. The group has established a goal of finding commercially viable ways to recycle or recover materials and energy from 25% of available PUR annually by 1995. Approximately 2/3 of annual PUR production ($\approx$3.3 billion pounds) is consumed by four industries: transportation, construction, furniture and bedding. It is estimated that 335 million pounds of PUR were recycled in 1989. Most of it came from the flexible foam (e.g., padding, cushions) segment of the industry.

3.7 Initial Activities in Automotive Plastics Recycling

In order to make the recovery of automobiles more feasible and reduce their landfilling, automotive companies are examining the recycling of plastic automotive parts. Recycling of plastic auto components is being studied at three different levels: a) through identification of plastic types and subsequent dismantlement for recycling; b) through solvent dissolution of thermoplastic mixtures from auto shredder residue; c) through pyrolizing of auto thermoset plastics for further use.

Identification of Plastic Auto Parts

To support the dismantlement and direct recycling of plastic auto parts, a voluntary coding system for identification of plastics used in automotive parts has been established by the Society of Automotive Engineers (SAE). The purpose of Recommended Practice SAE J1344 (March, 1988) is to provide information with regards to recycling as well as the selection of materials and procedures for the repairing and repainting of plastic parts. The recommended practice provides a listing of the plastic types used in automotive parts, the common trade name of a material, and the abbreviation which is to be imprinted or molded on the part. In addition, symbols for commercial plastic blends are also shown. The recommended practice can be obtained from SAE. The identification of plastic type used in automotive parts along with the design of modular plastic components would make ease of recovery simple and would allow for recycling of homogeneous plastic types with minimal contamination.

Solvent Dissolution of Plastic Auto Shredder Residue

Obsolete automobiles are typically shredded for the recovery of ferrous scrap. The nonmetallic fraction of shredded automobiles is auto shredder residue (ASR), a heterogeneous mixture of plastics, glass, rubber, fibers, foam, upholstery, dirt and fines. ASR may typically be contaminated with any of the original constituents of an automobile such as brake fluid, steering fluid, motor oil, gasoline or heavy metals. It is estimated that 250-275 lb. of plastic are used in a 3,500 lb. auto and that up to 1.2 million tons of waste plastic from ASR are generated per year [Bonsignore et al., 1991]. A breakdown of automotive plastic scrap for the 1981 model year has shown that the thermoplastics ABS, PUR, PP and PVC comprise 65% of waste plastic.

Mechanical separation and solvent dissolution of the thermoplastic portion of ASR is being studied at Argonne National Laboratories (ANL) as a way to recover additional resources from scrap autos and reduce landfill disposal of ASR. ASR samples from an auto shredder in the Chicago area were used in laboratory studies to identify the effectiveness of the mechanical cleaning and solvent dissolution. The mechanical cleaning portion, which also included removal of PUR foam, was accomplished using a vibrating screen separator. PUR foam chunks (greater than the initial screen size of 1") floated on top of the screen and was vacuumed off. The PUR was severely contaminated with residual oils. For the PUR foam scrap, an acetone wash followed by a water detergent wash removed most of the impregnating oils and entrapped dirt and fines. The remaining mixture of plastics (the "bottoms" from the 1" screen) was separated from ASR fines by passing the material over a 1/4" vibrating screen. The plastic "overs" mixture was washed with acetone at room temperature to remove oil, grease and mastics, and then it was subject to boiling ethylene dichloride (EDC). The EDC soluble plastics were recovered by solvent evaporation and yielded a blend of approximately 50% ABS and 50% PVC by weight [Bonsignore et al., 1991]. The soluble solids after EDC extraction were subject to xylene dissolution. Filtration and precipitation of the xylene solution yielded a high purity PP with some PE.

An initial charge of 2968 g of raw ASR was separated by vibrating screen into the following fractions: PUR foam (with contaminated oil), 275 g; plastic rich stream, 1906 g; and fines, 787 g. The composition of the resulting solvent cleaned products were as follows [Bonsignore et al., 1991]: cleaned PUR foam, 178 g; PVC/ABS composite, 180 g; PP, 54.5 g. Due to the size of the setup, no conclusions can yet be made on the effectiveness of such a system. Plans are underway to scale up to a 20 kg pilot lab size separation and solvent extraction system.

Pyrolyzing of Auto Thermosets

The recycle of thermosets (often phenolics) is being studied by automotive companies as a method of re-using the sheet molded compound (SMC) thermosets used to form automotive body panels. It is estimated that about 4% of the plastic waste disposed in landfills comes from scrapped automobiles [Vernyl, 1990]. Thermosets can be recycled by grinding the material into a fine powder and using it as an inert filler or by pyrolyzing it. Pyrolysis is the primary process being studied by the automotive industry as a method of handling scrap thermosets from automotive body part production. It consists of heating the thermosets in the absence or near absence of oxygen to drive off the volatiles present in the waste feed material. The resulting products are combustible gas or fuel oil which can be used to sustain the process, as feedstocks to the chemical industry, or as filler material in new product molding.

Additionally, the Council for Solid Waste Solutions has established a program to coordinate efforts for the recycling of durable plastic goods. The program will initially examine the technical aspects of durable goods recycling with emphasis on the disposal of automobile residue.

3.8 Sources of Plastic Recycling Information and Plastic Recycling Systems

Shown in Appendix B are sources of information on plastic recycling (publications, sources of markets, reports, and institutions and service organizations involved in plastic recycling).

In addition to the developing separation and cleaning systems discussed in Chapter 3, there are many cleaning and waste plastic processing vendors in the marketplace today. Most of the operations for cleaning and waste plastic processing are of a proprietary nature, and detailed information of a particular process may not be readily available to the public. However, the vendors should provide more detailed information for a specific case application and provide performance guarantees. Appendix C provides a listing of vendors in the plastics recycling field. Manufacturers of turnkey separation systems, granulators, shredders, cleaning, conveying, drying and processing equipment are also provided.

4. Buyers and Specifications for Waste Plastics

4.1 Buyers of Waste Plastic

Recovered plastic can be marketed for reuse in a number of ways:

- By directly dealing with a company which uses waste plastic in manufacturing
- By directly dealing with a plastic processor which will buy waste plastic and market the cleaned and decontaminated product
- By listing the recovered waste plastic in a waste exchange for marketing
- By marketing the recovered waste plastic through a scrap resin broker

Waste exchanges are typically sponsored by a state and provide a waste listing free of charge. The purpose of such a service is to serve as an information clearinghouse, directory and marketing facilitator so that waste materials may be reused or reprocessed in some manner. This often includes manufacturing by-product, surplus material, off-specification material, industrial waste and hazardous waste. Such exchanges are one route to marketing waste plastic. The waste exchange listing will typically include the waste product or desired product, the primary constituent(s), amount and frequency of generation, and the operation the material is a by-product from, if applicable. Samples are generally available upon request. Because of the extensive number of sources and types of plastics in waste, waste exchanges should generally be utilized only after other marketing methods (e.g., scrap resin brokers, plastic recycling companies) have been tried. The Illinois Environmental Protection Agency and the Illinois State Chamber of Commerce sponsor an Illinois based waste exchange (Industrial Material Exchange Service, 2200 Churchill Rd., #31, P.O. Box 19276, Springfield, IL, 62704-9276, phone (217) 782-0450). The waste exchanges in the U.S. and Canada are listed in Table 4.1.

A resin broker (a company which buys/sells off-grade resins, off-specification resins, regrind, obsolete or surplus virgin resins) or a plastic scrap handler (a company which may grind, clean, densify, pelletize, extrude, fabricate or process waste plastic in some way) is typically where plastic scrap is marketed after being collected at post consumer, post commercial, or industrial scrap level. Shipments in truckload quantities are typically preferred, but smaller loads down to bales are usually accepted with an accompanying reduction in price paid. According to the 1990-91 Directory of U.S. & Canadian Scrap Plastics Processors and Buyers, there are approximately 14 resin brokers

Table 4.1 U.S. and Canadian Waste Exchanges

Alberta Materials Waste Exchange
Industrial Development Department
Alberta Research Council
P.O. Box 8330, Postal Station F
Edmonton, Alberta
Canada T6H 5X2
(403) 450-5461

California Waste Exchange
Department od Health Services
Toxic Substances Control Division
714 P Street
Sacramento, CA 95814

Canadian Waste Materials Exchange
Ortech International
Sheridan Park Research Community
Mississauga, Ontario
Canada L5K 1B3
(916) 324-1807

Industrial Materials Exchange Service
P.O. Box 19276
2200 Churchill Road, #31
Springfield, IL 62794-9276
(217) 782-0450

Indiana Waste Exchange
Purdue University
School of Civil Engineering
West Lafayette, IN 47907
(317) 494-5036

Montana Industrial Waste Exchange
P.O. Box 1730
Helena, MT 59624
(406) 442-2405

Northeast Industrial Waste Exchange
90 Presidentail Plaza, Suite 122
Syracuse, NY 13202
(315) 422-6572

RENEW
Texas Water Commission
P.O. Box 13087
Austin, TX 78711
(512) 463-7773

Pacific Materials Exchange
South 3707 Godfrey Blvd.
Spokane, WA 99204
(509) 623-4244

Resource Exchange & News
400 Ann Street NW, Suite 301A
Grand Rapids, MI 49505
(616) 363-3262

Southeast Waste Exchange
Urban Institute, UNCC Station
Charlotte, NC 28223
(704) 547-2307

Southern Waste Info Exchange
P.O. Box 960
Tallahassee, FL 32302
(800) 441-7949

or scrap handlers in Illinois, with 11 in the Chicago metropolitan area, 2 in Joliet and 1 in Decatur. Five of the companies in metro Chicago are strictly brokers of plastic scrap. There are 20 additional brokers/processors in states neighboring Illinois: 8 in Wisconsin, 7 in Michigan, 2 in Iowa, 1 in Indiana, 1 in Kentucky and 1 in Missouri. The name, address and contact names of each handler/broker in Illinois and neighboring states along with the type of plastic each accepts is listed in Appendix A. The cross-listing addresses the following resins: ABS, Acetals, Acrylics, Engineering thermoplastics, HDPE, LDPE, Mixed thermoplastics, Nylons, PET, Polyolefins, PP, PS and PVC. While a majority of the brokers/handlers accept scrap plastic from manufacturers and processors, not all accept post-consumer material.

Additional information on markets for recycled plastics is available in the Illinois Recycled Materials Market Directory (ILENR/RR-87/01), and the accompanying update, available from the Illinois Department of Energy and Natural Resources, Office of Solid Waste and Renewable Resources, Springfield, Illinois. The State of Wisconsin Department of Natural Resources, Bureau of Solid Waste Management (Madison, WI.) also makes available a Wisconsin State Plastic Recycling Directory.

4.2 Specifications for Waste Plastic

Each of the above mentioned brokers/handlers generally have their own specifications regarding non-plastic contaminants and other plastic contaminants, and therefore should be consulted for acceptable levels. As may be expected, higher prices are paid for material with lesser amounts of contamination. All but a few require that plastic types be separated out from each other and pay more with clear/color sortation (as with HDPE bottles). A common request is that only baled material be shipped (rather than granulated) to allow for a final manual removal of contaminants. Although the price paid for granulated material is higher because of the cost of granulating, it is typically allowed by the buyer only after verification of cleanliness by inspecting incoming material in a baled form. Limits for non-plastic contaminants are typically:

- no metals
- <0.005% - <3% non-plastic
- must be clean

Limits for plastic contaminants are typically:

- <1% - <5% other plastic
- <1% color on clear/natural bottle loads
- no motor/vegetable oil bottles
- no PVC bottles

While 1% contamination of foreign material (other plastic or non-plastic) in an otherwise uniform load appears small, it can have a tremendous effect on the secondary plastic application. Foreign material on the order of 2 - 10% may be acceptable for plastic wood, but not for producing multilayer coextruded bottles. Rubbermaid, one of the largest U.S. purchasers of recycled plastic, has indicated that less than half of the waste plastic offered to the company is of a high enough quality for the company to recycle it into a new product. The cleanest material Rubbermaid receives is from integrated processors which grind, clean and produce pellets in one operation [Hill, 1990]. Table 4.2 shows a comparison between the physical properties of typical virgin and recycle composite grades of homopolymer and copolymer resin, and what a blow molder might specify in terms of melt index and resin density. Also shown in Table 4.2 are recycled material standards for properties of recycled clear and colored HDPE. Higher quality recycled plastic users also have difficulty with injection mold containers mixed in with blow molded containers because this results in manufacturing and product problems. Midwest Plastics, a company which produces HDPE piping from blow mold grade HDPE will produce a pipe with stress fractures if injection mold material is mixed into the batch.

One method of addressing plastic contamination in general is to link recycle product prices to product quality on a commonly accepted standardized system such as The American Society for Testing Materials (ASTM). The ASTM D-20 Committee addresses plastic recycling and degradable plastics. There are four draft standards regarding waste plastic contamination and recycling:

ASTM X-95-1-3	Guide for Development of Standards Relating to the Proper Use of Recycled Plastics (1/15/89)
ASTM X-95-2-6	Standard Practice for Generic Making of Plastic Products (4/15/90)
ASTM X-12-73	Proposed Standard Specification for Polyethylene Plastics Molding and Extrusion (7/20/90)
ASTM X-95-3-1	Proposed Standard Guide for Determining Visible Contaminant Content of Recycled Plastic Materials (9/24/90)

Table 4.2 Typical Physical Properties of Virgin and Recycled HDPE [a]

Parameter	Homopolymer		Copolymer	
	Virgin	Recycled Composite	Virgin	Recycled Composite
Melt Index (g/10 min.)	0.70	0.62	0.30	0.20-0.62
Density (g/cm^3)	0.96	0.96	0.954	0.965
Flexural modulus (psi)	219	223	186	189
Tensile strength [b] (psi)	4,290	4,340	3,840	4,020
Notched Izod Impact (J/cm)	1.3	1.5	1.3	0.9
Recycled Material Standards	*Unpigmented HDPE Bottles* [c]		*Colored HDPE Bottles* [d]	
Melt Index (g/10 min.)	0.5-1.0		0.2-0.5	
Density [e] (g/cm^3)	>0.958		$\leq$0.959	
Antioxidant added	Specify level		Specify level	
Tensile strength (psi)	$\geq$2,900		$\geq$2,030	
Secant modulus (psi)	$\geq$97,150		$\geq$89,900	
Typical Blow Molder Specifications [f]	*Homopolymer*		*Copolymer*	
Melt Index (g/10 min.)	0.5-0.9		0.3-0.8	
Density (g/cm^3)	0.958-0.965		0.947-0.955	

a. ASTM testing to be utilized for all properties specified.
b. Tensile strength at yield.
c. Unpigmented HDPE bottles are typically homopolymer.
d. Colored HDPE bottles are typically copolymer.
e. Colored plastics shall be corrected for colorant to reflect density of unpigmented base plastic.
f. Blow molder specifications shown for either virgin or recycled composite material.

In addition to developing commonly accepted ASTM standards for classifying recycled plastics, four quality grades of secondary plastics have been proposed by industry: superior grade - eligible for use in plastics packaging; high grade - eligible for use in structural quality; medium grade - eligible for use in higher end value added products such as flower pots, drain pipe, traffic cones, etc.; and low grade - for use in plastic lumber. The following quality testing to set differences between grades of material should be established [Rennie, 1990]:

- Surface contaminants - run tests on clean flake and specify the insolubles and inorganics allowed

- Soluble contaminants - run tests on clean flake indicating the amount of soluble product residue and degree of polymer degradation

- Polymer degradation - determine polymer degradation with respect to a virgin polymer and also determine the cross-contamination

The differences in the way products are molded also need to be addressed. This is particularly important with separated HDPE where loads may contain tubs such as soft drink base cups, deli containers, butter and yogurt which are injection molded, and bottles such as milk, water, juice, detergent and oil which are blow molded. The kind of HDPE that works best for blow molding is fractional melt resin, material that is stiff like taffy when melted for forming. The stiffness allows the material to be blown up like a balloon to form the shape of the inside of the blow mold. Injection molding material is runny like syrup when it is melted for forming. The runniness of injection molding material allows it to be forced into the injection mold under pressure which forms sharp corners and thin walls. The differently molded products are individually usable, but when mixed together they are not good for injection molding or blow molding due to the different melt indexes. To help resolve this, it has been proposed that the Society of the Plastics Industry numbering system for HDPE, which is coded as "2", be upgraded to "2-B" for blow molded containers and "2-I" for injection molded containers.

Appendix A: Plastic Scrap Handlers and Brokers

BEST ENTERPRISES INC. RESIN BROKER

3304 Commercial Ave.
Northbrook, IL 60062

PHONE: (708) 564-0400
FAX: (708) 564-1332

Best Enterprises Inc. buys/sells off-specification or off-grade resins, regrind and obsolete or surplus virgin resins.

RESINS:	FORM						SOURCES				PREFERRED SHIPMENT	
	WHOLE	GROUND	SHREDDED	BALED	FLAKED	REPROCESSED	PLASTICS PROCESSORS	OTHER MANUFACTURERS	POST COMMERCIAL	POST CONSUMER	TRUCK LOADS	RAIL CAR LOADS
ABS		✓				✓	✓	✓	✓	✓	✓	
ACETALS		✓				✓	✓	✓	✓		✓	
ACRYLIC		✓				✓	✓	✓	✓	✓	✓	
ENGINEERING THERMOPLASTICS		✓				✓	✓	✓	✓	✓	✓	
HDPE		✓	✓	✓		✓	✓	✓	✓	✓	✓	
LDPE		✓	✓	✓	✓	✓	✓	✓	✓	✓	✓	
MIXED THERMOPLASTICS		✓					✓	✓	✓	✓	✓	
NYLON		✓				✓	✓	✓	✓		✓	
PET		✓	✓	✓	✓	✓	✓	✓	✓	✓	✓	
POLYOLEFINS	✓	✓	✓	✓	✓	✓	✓	✓	✓	✓	✓	
PP	✓	✓	✓	✓	✓	✓	✓	✓	✓	✓	✓	
PS	✓	✓	✓	✓	✓	✓	✓	✓	✓	✓	✓	
PVC	✓	✓	✓	✓	✓	✓	✓	✓	✓	✓	✓	

152 Mixed Plastics Recycling Technology

B.M.W. PLASTICS CO.   **RESIN BROKER**

1515 N. Harlem Ave.   PHONE: (708) 848-8020
Oak Park, IL 60302   FAX: (708) 848-8071

B.M.W. Plastics Co. buys/sells off-specification or off-grade resins, regrind and obsolete or surplus virgin resins. They have been in operation for 18 years.

RESINS:	FORM							SOURCES				PREFERRED SHIPMENT	
	WHOLE	GROUND	SHREDDED	BALED	FLAKED	REPROCESSED	PLASTICS PROCESSORS	OTHER MANUFACTURERS	POST COMMERCIAL	POST CONSUMER	TRUCK LOADS	RAILCAR LOADS
ABS	■	■				■	■	■			■	
ACETALS	■	■				■	■	■				
ACRYLIC	■	■				■	■	■			■	
ENGINEERING THERMOPLASTICS	■	■				■	■	■				
HDPE	■	■				■	■	■			■	
LDPE	■	■				■	■	■			■	
MIXED THERMOPLASTICS												
NYLON	■	■				■	■	■				
PET		■				■	■	■			■	
POLYOLEFINS												
PP	■	■				■	■	■			■	
PS	■	■				■	■	■			■	
PVC (flexible)	■	■				■	■	■			■	

Appendix A: Plastic Scrap Handlers and Brokers

DLM AMERICAN PLASTICS

Larry Markin
570 Lake Cook Road Suite 207
Deerfield, IL 60015

RESIN BROKER
SCRAP HANDLER

PHONE: (708) 945-0300
FAX: (708) 945-0389

DLM American Plastics buys/sells off-specification or off-grade resins, regrind and obsolete or surplus virgin resins. They are brokers/distributors with warehousing in Chicago.

RESINS:	FORM						SOURCES			PREFERRED SHIPMENT	
	WHOLE	GROUND	SHREDDED	BALED	FLAKED	REPROCESSED	PLASTICS PROCESSORS	OTHER MANUFACTURERS	POST COMMERCIAL	POST CONSUMER	TRUCK LOADS	RAIL CAR LOADS
ABS	■	■	■	■	■	■	■	■	■		■	
ACETALS	■	■	■	■	■	■	■	■	■		■	
ACRYLIC	■	■	■	■	■	■	■	■	■		■	
ENGINEERING THERMOPLASTICS	■	■	■	■	■	■	■	■	■		■	
HDPE	■	■	■	■	■	■	■	■	■		■	
LDPE	■	■	■	■	■	■	■	■	■		■	
MIXED THERMOPLASTICS	■	■	■	■	■	■	■	■	■		■	
NYLON	■	■	■	■	■	■	■	■	■		■	
PET	□	□	□	□	□	□	□	□	□		□	
POLYOLEFINS	□	□	□	□	□	□	□	□	□		□	
PP	□	□	□	□	□	□	□	□	□		□	
PS	□	□	□	□	□	□	□	□	□		□	
PVC	□	□	□	□	□	□	□	□	□		□	

Mixed Plastics Recycling Technology

EAGLEBROOK PLASTICS, INC.

RESIN BROKER
SCRAP HANDLER

Eric Liewergen, Post Consumer Manager
2600 W. Roosevelt Road
Chicago, IL 60608

PHONE: (312) 638-2567
FAX: (312) 638-0006

Eaglebrook Plastics buys/sells off-specification or off-grade resins, regrind and obsolete or surplus virgin resins. They perform custom grinding and operate a cleaning or processing system handling post-consumer scrap. They make a product named "Durawood" from recycled plastic which serves as a lumber substitute, and a recycled pellet made from 100% post-consumer HDPE bottles that is suitable for blow-molding back into HIC containers.

RESINS:	FORM						SOURCES				PREFERRED SHIPMENT	
	WHOLE	GROUND	SHREDDED	BALED	FLAKED	REPROCESSED	PLASTICS PROCESSORS	OTHER MANUFACTURERS	POST COMMERCIAL	POST CONSUMER	TRUCK LOADS	RAILCAR LOADS
ABS	■	■				■	■	■			■	
ACETALS												
ACRYLIC												
ENGINEERING THERMOPLASTICS	■	■				■	■	■			■	
HDPE	■	■	■	■	■	■	■	■	■	■	■	
LDPE	■	■	■	■	■	■					■	
MIXED THERMOPLASTICS												
NYLON												
PET	■	■	■	■	■	■	■	■	■	■	■	
POLYOLEFINS												
PP	■	■	■	■	■	■	■	■			■	
PS	■	■	■	■	■	■	■	■			■	
PVC	■	■	■	■	■	■	■	■			■	

Appendix A: Plastic Scrap Handlers and Brokers

EMPIRE PLASTICS INC.   RESIN BROKER

Richard Keith, President
3352 Commercial Ave.
Northbrook, IL 60062   PHONE: (708) 564-8595

Empire Plastics buys/sells off-specification or off-grade resins, regrind and obsolete or surplus virgin resins. They specialize in engineering plastics and accomodate less than truck load accumulations (minimum 5,000 lbs.). The 12-year-old company services the entire continental U.S.

RESINS:	FORM						SOURCES				PREFERRED SHIPMENT	
	WHOLE	GROUND	SHREDDED	BALED	FLAKED	REPROCESSED	PLASTICS PROCESSORS	OTHER MANUFACTURERS	POST COMMERCIAL	POST CONSUMER	TRUCK LOADS	RAILCAR LOADS
ABS		X				X	X				X	
ACETALS		X				X	X				X	
ACRYLIC		X				X	X				X	
ENGINEERING THERMOPLASTICS		X				X	X				X	
HDPE		X				X	X				X	
LDPE		X				X	X				X	
MIXED THERMOPLASTICS												
NYLON		X				X	X				X	
PET												
POLYOLEFINS		X				X	X				X	
PP		X				X	X				X	
PS		X				X	X				X	
PVC		X				X	X				X	

FDA PLASTICS RECYCLING

Felix Akhimie, Plant Manager
2001 North 22nd St.
P.O. Box 966
Decatur, IL 62525

RESIN BROKER
SCRAP HANDLER

PHONE: (217) 429-3373

FDA Plastics Recycling buys/sells off-specification or off-grade resins, regrind, and obsolete or surplus resins. They perform custom grinding and operate a cleaning or processing system handling post-consumer scrap. The 9-year-old firm also buys industrial scrap.

RESINS:	FORM						SOURCES				PREFERRED SHIPMENT	
	WHOLE	GROUND	SHREDDED	BALED	FLAKED	REPROCESSED	PLASTICS PROCESSORS	OTHER MANUFACTURERS	POST COMMERCIAL	POST CONSUMER	TRUCK LOADS	RAILCAR LOADS
ABS	☑	☑	☑	☑	☑	☑	☑	☑	☑	☑	☑	
ACETALS												
ACRYLIC	☑	☑	☑	☑	☑	☑	☑	☑	☑	☑	☑	
ENGINEERING THERMOPLASTICS	☑	☑	☑	☑	☑	☑	☑	☑	☑	☑	☑	
HDPE	☑	☑	☑	☑	☑	☑	☑	☑	☑	☑	☑	
LDPE	☑	☑	☑	☑	☑	☑	☑	☑	☑	☑	☑	
MIXED THERMOPLASTICS												
NYLON												
PET	☑	☑	☑	☑	☑	☑	☑	☑	☑	☑	☑	
POLYOLEFINS												
PP	☑	☑	☑	☑	☑	☑	☑	☑	☑	☑	☑	
PS	☑	☑	☑	☑	☑	☑	☑	☑	☑	☑	☑	
PVC	☑	☑	☑	☑	☑	☑	☑	☑	☑	☑	☑	

Appendix A: Plastic Scrap Handlers and Brokers

JLM PLASTICS CORPORATION

Joe Mitchell, Operations Manager
1012 Collins St.
Joliet, IL 60432

RESIN BROKER
SCRAP HANDLER

PHONE: (815) 722-0066
FAX: (815) 722-0535

JLM Plastics buys/sells off-specification or off-grade resins, regrind and obsolete or surplus virgin resins. They perform custom grinding. Since 1979, JLM has marketed high quality regrind materials. They maintain a large inventory of commodity and engineering materials.

RESINS:	FORM						SOURCES				PREFERRED SHIPMENT	
	WHOLE	GROUND	SHREDDED	BALED	FLAKED	REPROCESSED	PLASTICS PROCESSORS	OTHER MANUFACTURERS	POST COMMERCIAL	POST CONSUMER	TRUCK LOADS	RAIL CAR LOADS
ABS	■	■	■		■	■	■	■	■		■	
ACETALS	■	■	■		■	■	■	■	■		■	
ACRYLIC	■	■	■	■	■	■	■	■	■		■	
ENGINEERING THERMOPLASTICS	■	■	■		■	■	■	■	■		■	
HDPE	■	■	■		■	■	■	■	■		■	
LDPE	■	■	■	■	■	■	■	■	■		■	
MIXED THERMOPLASTICS												
NYLON	■	■	■		■	■	■	■	■		■	
PET	■	■	■		■	■	■	■	■		■	
POLYOLEFINS	■	■	■	■	■	■	■	■	■		■	
PP	■	■	■		■	■	■	■	■		■	
PS	■	■	■	■	■	■	■	■	■		■	
PVC	■	■	■	■	■	■	■	■	■		■	

158 Mixed Plastics Recycling Technology

MID CONTINENT PLASTICS INC.   **RESIN BROKER**

Jack Blitvich, President
6401 W. 65th St.
Bedford Park, IL 60638

PHONE: (708) 496-3232
FAX: (708) 496-3237

Mid Continent Plastics buys/sells off-specification or off-grade resins, regrind and obsolete or surplus virgin resins. They buy and sell all types of thermoplastic resins.

RESINS:	FORM							SOURCES				PREFERRED SHIPMENT	
	WHOLE	GROUND	SHREDDED	BALED	FLAKED	REPROCESSED	PLASTICS PROCESSORS	OTHER MANUFACTURERS	POST COMMERCIAL	POST CONSUMER	TRUCK LOADS	RAILCAR LOADS
ABS	■	■				■	■	■			■	
ACETALS	■	■				■	■	■			■	
ACRYLIC	■	■				■	■	■			■	
ENGINEERING THERMOPLASTICS	■	■				■	■	■			■	
HDPE	■	■				■	■	■			■	
LDPE	■	■				■	■	■			■	
MIXED THERMOPLASTICS												
NYLON	■	■				■	■	■			■	
PET												
POLYOLEFINS	■	■				■	■	■			■	
PP	■	■				■	■	■			■	
PS	■	■				■	■	■			■	
PVC												

Appendix A: Plastic Scrap Handlers and Brokers

MIDWEST RECYCLING COMPANY, INC.

Mr. Rudnick, President
1086 Old Elm
Glencoe, IL 60022

RESIN BROKER
SCRAP HANDLER

PHONE: (312) 835-2020

Midwest Recycling Company buys/sells off-specification or off-grade resins, regrind and obsolete or surplus virgin resins. They also perform custom grinding. The 17-year-old firm performs toll work pelletizing.

RESINS:	FORM						SOURCES				PREFERRED SHIPMENT	
	WHOLE	GROUND	SHREDDED	BALED	FLAKED	REPROCESSED	PLASTICS PROCESSORS	OTHER MANUFACTURERS	POST COMMERCIAL	POST CONSUMER	TRUCK LOADS	RAILCAR LOADS
ABS	☒	☒					☒	☒	☒		☒	
ACETALS												
ACRYLIC	☒	☒					☒	☒	☒		☒	
ENGINEERING THERMOPLASTICS	☒	☒					☒	☒	☒		☒	
HDPE	☒	☒		☒		☒	☒	☒	☒		☒	
LDPE	☒	☒		☒			☒	☒	☒		☒	
MIXED THERMOPLASTICS	☒	☒		☒		☒	☒	☒	☒		☒	
NYLON				☒			☒	☒	☒		☒	
PET				☒			☒	☒	☒		☒	
POLYOLEFINS	☒	☒				☒	☒	☒	☒		☒	
PP				☒			☒	☒	☒		☒	
PS	☒	☒					☒	☒	☒		☒	
PVC	☒	☒					☒	☒	☒		☒	

MRC POLYMERS INC.

Daniel Eberhardt, President
George Staniulis, Sales & Product Development
1716 W. Webster
Chicago, IL 60614

SCRAP HANDLER

PHONE: (312) 276-6345
FAX: (312) 276-4431

MRC Polymers occasionally performs custom grinding.

RESINS:	FORM						SOURCES				PREFERRED SHIPMENT	
	WHOLE	GROUND	SHREDDED	BALED	FLAKED	REPROCESSED	PLASTICS PROCESSORS	OTHER MANUFACTURERS	POST COMMERCIAL	POST CONSUMER	TRUCK LOADS	RAILCAR LOADS
ABS												
ACETALS												
ACRYLIC												
ENGINEERING THERMOPLASTICS	■	■	■	■	■			■	■		■	
HDPE												
LDPE												
MIXED THERMOPLASTICS												
NYLON	□	□	□	□	□			□	□			
PET				■					■	■		
POLYOLEFINS												
PP												
PS												
PVC												

Appendix A: Plastic Scrap Handlers and Brokers

NUCON CORPORATION

Peter Pigott, President
540 Frontage Rd.
Northfield, IL 60093

RESIN BROKER
SCRAP HANDLER

PHONE: (708) 446-6777
FAX: (708) 446-6826
TELEX: 754 549 UD

Nucon Corporation buys off-specification or off-grade resins, regrind and obsolete or surplus virgin resins. They operate a cleaning or processing system handling post-consumer scrap and make recycled plastic pallets.

RESINS:	FORM						SOURCES				PREFERRED SHIPMENT	
	WHOLE	GROUND	SHREDDED	BALED	FLAKED	REPROCESSED	PLASTICS PROCESSORS	OTHER MANUFACTURERS	POST COMMERCIAL	POST CONSUMER	TRUCK LOADS	RAILCAR LOADS
ABS												
ACETALS												
ACRYLIC												
ENGINEERING THERMOPLASTICS		☐				☐	☐	☐	☐	☐	☐	
HDPE		☐		☐	☐	☐	☐	☐	☐	☐	☐	
LDPE		☐		☐		☐	☐	☐	☐	☐	☐	
MIXED THERMOPLASTICS												
NYLON												
PET		☐		☐	☐	☐	☐	☐	☐	☐	☐	
POLYOLEFINS												
PP		☐		☐	☐	☐	☐	☐	☐	☐	☐	
PS		☐		☐	☐	☐	☐	☐	☐	☐	☐	
PVC												

PLASTIC MATERIALS UNLIMITED

RESIN BROKER

Paul Wehrwein, President
4129 White Ash Rd.
P.O. Box 512
Crystal Lake, IL 60014

PHONE: (815) 455-5083
FAX: (815) 459-6786
ACCT # 3003

Plastic Materials Unlimited buys/sells off-specification or off-grade resins, regrind and obsolete or surplus virgin resins.

RESINS:	FORM						SOURCES				PREFERRED SHIPMENT	
	WHOLE	GROUND	SHREDDED	BALED	FLAKED	REPROCESSED	PLASTICS PROCESSORS	OTHER MANUFACTURERS	POST COMMERCIAL	POST CONSUMER	TRUCK LOADS	RAIL CAR LOADS
ABS	☐	☐				☐	☐		☐	☐	☐	
ACETALS		☐								☐	☐	
ACRYLIC	☐	☐							☐	☐	☐	
ENGINEERING THERMOPLASTICS		☐				☐				☐	☐	
HDPE	☐	☐		☐		☐	☐		☐	☐	☐	
LDPE	☐	☐		☐		☐	☐		☐	☐	☐	
MIXED THERMOPLASTICS												
NYLON		☐								☐	☐	
PET		☐		☐					☐	☐	☐	
POLYOLEFINS	☐	☐		☐		☐	☐		☐	☐	☐	
PP	☐	☐		☐		☐	☐		☐	☐	☐	
PS	☐	☐		☐		☐	☐		☐	☐	☐	
PVC	☐	☐		☐		☐	☐		☐	☐	☐	

Appendix A: Plastic Scrap Handlers and Brokers   163

POLY PRO PRODUCTS INC.

Charles Ward, President
711 Benton Ave.   P.O. Box 69
Joliet, IL   Thornton, IL 60476

RESIN BROKER
SCRAP HANDLER

PHONE: (815) 727-3739

Poly Pro Products buys off-grade resins, buys/sells regrind and performs custom grinding. They operate a cleaning or processing system handling post-consumer scrap. The firm has been in operation more than 12 years.

RESINS:	FORM						SOURCES				PREFERRED SHIPMENT	
	WHOLE	GROUND	SHREDDED	BALED	FLAKED	REPROCESSED	PLASTICS PROCESSORS	OTHER MANUFACTURERS	POST COMMERCIAL	POST CONSUMER	TRUCK LOADS	RAILCAR LOADS
ABS												
ACETALS												
ACRYLIC												
ENGINEERING THERMOPLASTICS												
HDPE	☐	☐	☐	☐	☐	☐	☐	☐	☐	☐	☐	
LDPE	☐	☐	☐	☐	☐	☐	☐	☐	☐	☐	☐	
MIXED THERMOPLASTICS												
NYLON												
PET												
POLYOLEFINS												
PP												
PS												
PVC												

S.B. RECYCLERS

Ronald Bajarunas, President
1501 E. 142nd St.
Dolton, IL 60419

P.O. Box 591
Midlothian, IL 60445

**RESIN BROKER
SCRAP HANDLER**

PHONE: (312) 841-3800

S.B. Recyclers buys/sells off-specification or off-grade resins, regrind and obsolete or surplus virgin resins. They perform custom grinding.

RESINS:	FORM						SOURCES				PREFERRED SHIPMENT	
	WHOLE	GROUND	SHREDDED	BALED	FLAKED	REPROCESSED	PLASTICS PROCESSORS	OTHER MANUFACTURERS	POST COMMERCIAL	POST CONSUMER	TRUCK LOADS	RAIL CAR LOADS
ABS	☒	☒				☒	☒	☒	☒		☒	
ACETALS												
ACRYLIC												
ENGINEERING THERMOPLASTICS												
HDPE	☒	☒	☒		☒	☒	☒		☒	☒	☒	
LDPE	☒	☒	☒	☒	☒	☒	☒	☒	☒		☒	
MIXED THERMOPLASTICS												
NYLON												
PET												
POLYOLEFINS												
PP	☒	☒	☒	☒	☒	☒	☒	☒	☒		☒	
PS	☒	☒	☒	☒	☒	☒	☒	☒	☒		☒	
PVC												

Appendix A: Plastic Scrap Handlers and Brokers

THE HESS COMPANY INC. (T.H.C.)

Jeff Hess, President
Justin Hess, Vice President
3717 E. 82nd Ct.
Merrillville, IN 46410

RESIN BROKER
SCRAP HANDLER

PHONE: (219) 942-2748

The Hess Company sells regrind and performs custom grinding. They have been in operation since 1978.

RESINS:	FORM						SOURCES				PREFERRED SHIPMENT	
	WHOLE	GROUND	SHREDDED	BALED	FLAKED	REPROCESSED	PLASTICS PROCESSORS	OTHER MANUFACTURERS	POST COMMERCIAL	POST CONSUMER	TRUCK LOADS	RAIL CAR LOADS
ABS												
ACETALS												
ACRYLIC												
ENGINEERING THERMOPLASTICS												
HDPE	☐			☐				☐	☐	☐	☐	
LDPE												
MIXED THERMOPLASTICS												
NYLON												
PET	☐			☐				☐	☐	☐	☐	
POLYOLEFINS												
PP												
PS												
PVC	☐			☐				☐	☐	☐	☐	

HAMMER'S PLASTIC RECYCLING CORP.

Floyd V. Hammer, President
Hwy 20 and 65 North
Iowa Falls, IA 50126

RESIN BROKER
SCRAP HANDLER

PHONE: (515) 648-5073
FAX: (515) 648-5074
TELEX: 883413 UNUD

Hammer's Plastic Recycling Corp. buys off-specification or off-grade resins and obsolete or surplus virgin resins. Formed in 1984, Hammer's Plastic Recycling manufactures products made from 100% recycled mixed plastic wastes.

RESINS:	FORM						SOURCES				PREFERRED SHIPMENT	
	WHOLE	GROUND	SHREDDED	BALED	FLAKED	REPROCESSED	PLASTICS PROCESSORS	OTHER MANUFACTURERS	POST COMMERCIAL	POST CONSUMER	TRUCK LOADS	RAIL CAR LOADS
ABS												
ACETALS												
ACRYLIC												
ENGINEERING THERMOPLASTICS												
HDPE	☐			☐						☐	☐	
LDPE	☐			☐			☐	☐			☐	
MIXED THERMOPLASTICS	☐			☐			☐	☐			☐	
NYLON												
PET	☐			☐						☐	☐	
POLYOLEFINS	☐			☐			☐	☐			☐	
PP	☐			☐			☐	☐			☐	
PS												
PVC												

Appendix A: Plastic Scrap Handlers and Brokers

NCS PLASTICS INC.
Verlin Eaker, Marketing Director
1210 9th St. S.W.
Cedar Rapids, IA 52404

RESIN BROKER
SCRAP HANDLER
PHONE: (319) 363-2112
FAX: (319) 363-2111

NCS Plastics Inc. sells regrind and operates a cleaning or processing system handling post-consumer scrap.

RESINS:	FORM						SOURCES				PREFERRED SHIPMENT	
	WHOLE	GROUND	SHREDDED	BALED	FLAKED	REPROCESSED	PLASTICS PROCESSORS	OTHER MANUFACTURERS	POST COMMERCIAL	POST CONSUMER	TRUCK LOADS	RAILCAR LOADS
ABS												
ACETALS												
ACRYLIC												
ENGINEERING THERMOPLASTICS												
HDPE	☐			☐						☐	☐	
LDPE	☐			☐						☐	☐	
MIXED THERMOPLASTICS												
NYLON												
PET	☐			☐						☐	☐	
POLYOLEFINS												
PP												
PS												
PVC												

SUPERIOR FIBER COMPANY, INC.

Ted Zdobylak, President
2201 Crosshill Road
Louisville, KY 40206

RESIN BROKER
SCRAP HANDLER

PHONE: (502) 895-7059

Superior Fiber Company buys/sells off-specification or off-grade resins, regrind and obsolete or surplus virgin resins. They operate a cleaning or processing system handling post-consumer scrap.

RESINS:	FORM						SOURCES				PREFERRED SHIPMENT	
	WHOLE	GROUND	SHREDDED	BALED	FLAKED	REPROCESSED	PLASTICS PROCESSORS	OTHER MANUFACTURERS	POST COMMERCIAL	POST CONSUMER	TRUCK LOADS	RAILCAR LOADS
ABS	☐	☐	☐	☐	☐	☐	☐	☐	☐	☐	☐	☐
ACETALS												
ACRYLIC												
ENGINEERING THERMOPLASTICS												
HDPE	☐	☐	☐	☐	☐	☐	☐	☐	☐	☐	☐	☐
LDPE	☐	☐	☐	☐	☐	☐	☐	☐	☐	☐	☐	☐
MIXED THERMOPLASTICS												
NYLON												
PET												
POLYOLEFINS												
PP	☐	☐	☐	☐	☐	☐	☐	☐	☐	☐	☐	☐
PS	☐	☐	☐	☐	☐	☐	☐	☐	☐	☐	☐	☐
PVC	☐	☐	☐	☐	☐	☐	☐	☐	☐	☐	☐	☐

Appendix A: Plastic Scrap Handlers and Brokers

AMERICAN COMMODITIES, INC. RESIN BROKER

Mark Lieberman, President
16165 W. 12 Mile Rd.
Southfield, MI 48076

PHONE: (313) 559-5300
FAX: (313) 559-0020

American Commodities, Inc. was established in 1984 and is also located in Tampa, Florida; Flint, Michigan; and in Ontario, California. They buy/sell off-specification or off-grade resins, regrind and obsolete or surplus virgin resins. They perform custom grinding, compounding, blending, coloring and repelletizing.

RESINS:	FORM						SOURCES				PREFERRED SHIPMENT	
	WHOLE	GROUND	SHREDDED	BALED	FLAKED	REPROCESSED	PLASTICS PROCESSORS	OTHER MANUFACTURERS	POST COMMERCIAL	POST CONSUMER	TRUCK LOADS	RAIL CAR LOADS
ABS	☐	☐				☐	☐	☐	☐		☐	☐
ACETALS	☐	☐				☐	☐	☐	☐		☐	
ACRYLIC	☐	☐	☐			☐	☐	☐	☐	☐	☐	
ENGINEERING THERMOPLASTICS	☐	☐				☐	☐	☐	☐		☐	
HDPE	☐	☐	☐	☐	☐	☐	☐	☐	☐	☐	☐	☐
LDPE	☐	☐		☐		☐	☐	☐	☐		☐	☐
MIXED THERMOPLASTICS												
NYLON	☐	☐			☐	☐	☐	☐	☐		☐	
PET	☐	☐		☐	☐	☐	☐	☐	☐	☐	☐	
POLYOLEFINS	☐	☐				☐	☐	☐	☐		☐	
PP	☐	☐				☐	☐	☐	☐		☐	
PS	☐	☐			☐	☐	☐	☐	☐		☐	☐
PVC	☐	☐				☐	☐	☐	☐		☐	

AMERICAN PLASTICS RECYCLING GROUP

Terrence L. Blakely, President
1790 E. Bluewater Hwy.
P.O. Box 68
Ionia, MI 48846

RESIN BROKER
SCRAP HANDLER

PHONE: (616) 527-6677

Processed Plastics buys off-specification or off-grade resins, regrind and obsolete or surplus virgin resins. They perform custom grinding. They make plastic lumber, park benches, car stops and sheets. They process commingled, contaminated material by low pressure extrusion molding.

RESINS:	FORM						SOURCES				PREFERRED SHIPMENT	
	WHOLE	GROUND	SHREDDED	BALED	FLAKED	REPROCESSED	PLASTICS PROCESSORS	OTHER MANUFACTURERS	POST COMMERCIAL	POST CONSUMER	TRUCK LOADS	RAIL CAR LOADS
ABS												
ACETALS												
ACRYLIC												
ENGINEERING THERMOPLASTICS												
HDPE	■	■	■	■	■	■	■	■	■	■	■	
LDPE	■	■	■	□	□	□	□	□	■	■	□	
MIXED THERMOPLASTICS	□	□	□	■	□	□	□	■	■	■	□	
NYLON												
PET	□	□	□	□	□	□	□	□	□	□	□	
POLYOLEFINS	□	□	□	□	□	□	□	■	□	■		
PP	□	□	□	□	□	□	□	□	□	□	□	
PS	□	□	□	□	□	□	□	■	□	□	□	
PVC												

Appendix A: Plastic Scrap Handlers and Brokers

MODERN MACHINERY OF BEAVERTON INC.

Larry Richardson, President
3031 Guernsey Rd.
Beaverton, MI 48612

RESIN BROKER
SCRAP HANDLER

PHONE: (517) 435-9071
FAX: (517) 435-3940

Modern Machinery of Beaverton operates a cleaning system which they designed and constructed and which does not involve the use of hazardous ingredients. They also buy/sell regrind.

RESINS:	FORM						SOURCES				PREFERRED SHIPMENT	
	WHOLE	GROUND	SHREDDED	BALED	FLAKED	REPROCESSED	PLASTICS PROCESSORS	OTHER MANUFACTURERS	POST COMMERCIAL	POST CONSUMER	TRUCK LOADS	RAILCAR LOADS
ABS												
ACETALS												
ACRYLIC												
ENGINEERING THERMOPLASTICS												
HDPE		☐			☐		☐	☐	☐	☐	☐	
LDPE												
MIXED THERMOPLASTICS		☐			☐		☐	☐	☐	☐	☐	
NYLON												
PET		☐			☐		☐	☐	☐	☐	☐	
POLYOLEFINS		☐			☐		☐	☐	☐	☐	☐	
PP		☐			☐		☐	☐	☐	☐	☐	
PS		☐			☐		☐	☐	☐	☐	☐	
PVC		☐			☐		☐	☐	☐	☐	☐	

NAPCO (NATIONAL PLASTIC CONVERTERS, INC.)

Napoleon G. Sharma, President
P.O. Box 2114
Southfield, MI 48037-2114

RESIN BROKER
SCRAP HANDLER

PHONE: (313) 352-1120
FAX: (313) 352-3277

NAPCO expects to be operative by October, 1990 with initial production capacity of 5.5 millions pounds monthly. They will obtain scrap plastics from the waste-stream and collectors and have an expected work force of 92 workers. They buy off-specification or off-grade resins, regrind and obsolete or surplus virgin resins. They perform custom grinding and make identifiable recycled pellets.

RESINS:	FORM						SOURCES				PREFERRED SHIPMENT	
	WHOLE	GROUND	SHREDDED	BALED	FLAKED	REPROCESSED	PLASTICS PROCESSORS	OTHER MANUFACTURERS	POST COMMERCIAL	POST CONSUMER	TRUCK LOADS	RAILCAR LOADS
ABS	☐						☐	☐	☐	☐	☐	
ACETALS												
ACRYLIC												
ENGINEERING THERMOPLASTICS												
HDPE	☐						☐	☐	☐	☐	☐	
LDPE	☐						☐	☐	☐	☐	☐	
MIXED THERMOPLASTICS												
NYLON												
PET												
POLYOLEFINS												
PP	☐						☐	☐	☐	☐	☐	
PS	☐						☐	☐	☐	☐	☐	
PVC	☐						☐	☐	☐	☐	☐	

Appendix A: Plastic Scrap Handlers and Brokers

PLASTIC CENTRAL

Bill Neumann, Broker
8438 N. 12th St.
Kalamazoo, MI 49009

RESIN BROKER
SCRAP HANDLER

PHONE: (616) 381-6302
FAX: (616) 381-6302

Plastic Central buys/sells off-specification or off-grade resins, regrind and obsolete or surplus virgin resins. They perform custom grinding and deal with export and domestic markets.

RESINS:	FORM						SOURCES				PREFERRED SHIPMENT	
	WHOLE	GROUND	SHREDDED	BALED	FLAKED	REPROCESSED	PLASTICS PROCESSORS	OTHER MANUFACTURERS	POST COMMERCIAL	POST CONSUMER	TRUCK LOADS	RAILCAR LOADS
ABS	☐	■	■	■	■	■	■	■	■	■	■	☐
ACETALS												
ACRYLIC	☐	☐	■	■	☐	☐	☐	☐	☐	■	☐	☐
ENGINEERING THERMOPLASTICS												
HDPE	☐	☐	■	■	■	☐	☐	■	☐	☐	☐	
LDPE	☐	☐	☐	■	☐	☐	☐	☐	☐	☐	☐	
MIXED THERMOPLASTICS												
NYLON	☐	☐	☐	■	☐	☐	☐	☐	☐	☐	☐	
PET	☐	☐	■	☐	☐	☐	☐	☐	☐	☐	☐	
POLYOLEFINS												
PP	☐	☐	☐	■	☐	☐	☐	☐	☐	☐	☐	
PS	☐	☐	☐	☐	☐	☐	☐	■	☐	☐	☐	
PVC	☐	☐	☐	☐	☐	☐	☐	■	☐	■	■	

SECONDARY POLYMERS, LTD.

Stuart Kay, President
5151 Bellevue
Mail: 10040 Freeland
Detroit, MI 48227

**RESIN BROKER
SCRAP HANDLER**

PHONES: (313) 491-8500
(313) 922-7444
FAX: (313) 933-7710

Secondary Polymers, Ltd. sells regrind and performs custom grinding. A major part of their business is the operation of a cleaning or processing system handling post-consumer scrap.

RESINS:	FORM						SOURCES				PREFERRED SHIPMENT	
	WHOLE	GROUND	SHREDDED	BALED	FLAKED	REPROCESSED	PLASTICS PROCESSORS	OTHER MANUFACTURERS	POST COMMERCIAL	POST CONSUMER	TRUCK LOADS	RAILCAR LOADS
ABS												
ACETALS												
ACRYLIC												
ENGINEERING THERMOPLASTICS												
HDPE	☐				☐				☐	☐	☐	
LDPE												
MIXED THERMOPLASTICS												
NYLON												
PET	☐				☐							
POLYOLEFINS												
PP	☐				☐				☐	☐	☐	
PS	☐				☐				☐	☐	☐	
PVC	☐				☐				☐	☐	☐	

Appendix A: Plastic Scrap Handlers and Brokers

UNIPLAS, INC.

Donald C. Root, President
2769 Hubert Road
Brighton, MI 48116

RESIN BROKER
SCRAP HANDLER

PHONE: (313) 229-2840

Uniplas buys/sells off-specification or off-grade resins, regrind and obsolete or surplus virgin resins. They source, grind, reprocess and compound materials.

RESINS:	FORM						SOURCES				PREFERRED SHIPMENT	
	WHOLE	GROUND	SHREDDED	BALED	FLAKED	REPROCESSED	PLASTICS PROCESSORS	OTHER MANUFACTURERS	POST COMMERCIAL	POST CONSUMER	TRUCK LOADS	RAILCAR LOADS
ABS	☒	☒	☒			☒	☒	☒	☒	☒	☒	
ACETALS		☒				☒	☒	☒	☒	☒		
ACRYLIC		☒				☒	☒	☒	☒	☒	☒	
ENGINEERING THERMOPLASTICS		☒				☒	☒	☒	☒	☒		
HDPE	☒	☒	☒	☒	☒	☒	☒	☒	☒	☒	☒	
LDPE	☒	☒	☒	☒	☒	☒	☒	☒	☒	☒	☒	
MIXED THERMOPLASTICS												
NYLON		☒				☒	☒	☒	☒	☒	☒	
PET												
POLYOLEFINS		☒					☒	☒	☒	☒	☒	
PP	☒	☒	☒	☒	☒	☒	☒	☒	☒	☒	☒	
PS	☒	☒				☒	☒	☒	☒	☒		
PVC												

KOLLER CRAFT PLASTIC PRODUCTS

RESIN BROKER

A.J. Koller Jr., President
1400 South Highway 141
P.O. Box K
Fenton, MO 63026

PHONE: (314) 343-9220
FAX: (314) 343-1034

Koller Craft Plastic Products buys/sells off-specification or off-grade resins, regrind and obsolete or surplus virgin resins. They have been in business since 1941 and employ 70 people. They own a custom injection molder used primarily for large parts.

RESINS:	FORM						SOURCES				PREFERRED SHIPMENT	
	WHOLE	GROUND	SHREDDED	BALED	FLAKED	REPROCESSED	PLASTICS PROCESSORS	OTHER MANUFACTURERS	POST COMMERCIAL	POST CONSUMER	TRUCK LOADS	RAILCAR LOADS
ABS												
ACETALS												
ACRYLIC												
ENGINEERING THERMOPLASTICS												
HDPE		☒	☒			☐	☐			☐	☐	
LDPE												
MIXED THERMOPLASTICS												
NYLON												
PET												
POLYOLEFINS												
PP		☒	☒			☐	☐			☐	☐	
PS												
PVC												

Appendix A: Plastic Scrap Handlers and Brokers

A TO Z RECYCLING

James F. Cherney, Recycling Coordinator
P.O. Box 675
Marshfield, WI 54449

RESIN BROKER
SCRAP HANDLER

PHONE: (715) 384-9308

A to Z Recycling accepts clean garbage bags, shopping bags, film, clear stretch wrap and clear or yellow insulation bags. They are seeking buyers of vinyl and PVC turnings.

RESINS:	FORM						SOURCES				PREFERRED SHIPMENT	
	WHOLE	GROUND	SHREDDED	BALED	FLAKED	REPROCESSED	PLASTICS PROCESSORS	OTHER MANUFACTURERS	POST COMMERCIAL	POST CONSUMER	TRUCK LOADS	RAILCAR LOADS
ABS												
ACETALS												
ACRYLIC												
ENGINEERING THERMOPLASTICS												
HDPE		■					■		■	■		
LDPE	■	■					■		■	■		
MIXED THERMOPLASTICS												
NYLON												
PET	■	■					■			■		
POLYOLEFINS												
PP												
PS												
PVC		■	■									

CAPITOL POLYMERS INC.

C. P. Jahn, President
1422 Packers Ave.
Madison, WI 53704

RESIN BROKER
SCRAP HANDLER

PHONE: (608) 846-9310
FAX: (608) 846-2748

Capitol Polymers Inc. performs custom pelletizing on a toll basis. The company now reprocesses post-industrial and post-commercial scrap plastics with future plans to enter the post consumer scrap market.

RESINS:	FORM						SOURCES				PREFERRED SHIPMENT	
	WHOLE	GROUND	SHREDDED	BALED	FLAKED	REPROCESSED	PLASTICS PROCESSORS	OTHER MANUFACTURERS	POST COMMERCIAL	POST CONSUMER	TRUCK LOADS	RAILCAR LOADS
ABS												
ACETALS												
ACRYLIC												
ENGINEERING THERMOPLASTICS												
HDPE	■	■	■	■	■		■	■	■		■	■
LDPE	■	■	■	■	■		■	■	■		■	■
MIXED THERMOPLASTICS												
NYLON												
PET												
POLYOLEFINS												
PP	■	■	■	■	■		■	■	■		■	■
PS	■	■	■	■	■		■	■	■		■	■
PVC												

Appendix A: Plastic Scrap Handlers and Brokers

PLASTIC RECOVERY SERVICE

Jeff Kernen, President
118 Oakridge Dr.
North Prairie, WI 53153

RESIN BROKER
SCRAP HANDLER

PHONE: (414) 392-3515

Plastic Recovery Service buys/sells off-specification or off-grade resins, regrind and obsolete or surplus virgin resins. They do custom grinding. They are industrial plastic recyclers, dealing with high volumes.

RESINS:	FORM						SOURCES				PREFERRED SHIPMENT	
	WHOLE	GROUND	SHREDDED	BALED	FLAKED	REPROCESSED	PLASTICS PROCESSORS	OTHER MANUFACTURERS	POST COMMERCIAL	POST CONSUMER	TRUCK LOADS	RAIL CAR LOADS
ABS	☐	☐	☐	☐	☐	☐			☐		☐	☐
ACETALS									☐			
ACRYLIC	☐	☐	☐						☐		☐	☐
ENGINEERING THERMOPLASTICS	☐	☐	☐						☐		☐	☐
HDPE	☐	☐	☐	☐	☐	☐			☐	☐	☐	☐
LDPE	☐	☐	☐	☐	☐	☐			☐	☐	☐	☐
MIXED THERMOPLASTICS												
NYLON												
PET												
POLYOLEFINS												
PP	☐	☐	☐	☐	☐	☐			☐		☐	☐
PS	☐	☐	☐	☐	☐	☐			☐		☐	☐
PVC	☐	☐	☐	☐	☐	☐			☐		☐	☐

POLY-ANNA PLASTIC PRODUCTS INC.

Marty Forman, President
Curt Leffingwell, Director of Operations
6860 N. Teutonia Ave.
Mequon, WI 53209

RESIN BROKER
SCRAP HANDLER

PHONE: (414) 351-5990
FAX: (414) 351-3443

Poly-Anna Plastic Products buys/sells off-specification or off-grade resins, regrind and obsolete or surplus virgin resins. They perform custom grinding and operate a cleaning or processing system that handles post-consumer scrap. They make recycled plastic custom injection molding. This firm is Wisconsin's only total approach plastic recycler of industrial, commercial and post-consumer plastics including difficult to recycle foam polystyrene, vinyl, mixed PET, foam PE, etc.

RESINS:	FORM						SOURCES				PREFERRED SHIPMENT	
	WHOLE	GROUND	SHREDDED	BALED	FLAKED	REPROCESSED	PLASTICS PROCESSORS	OTHER MANUFACTURERS	POST COMMERCIAL	POST CONSUMER	TRUCK LOADS	RAILCAR LOADS
ABS	■						■	■	■		■	
ACETALS												
ACRYLIC												
ENGINEERING THERMOPLASTICS	■	■		■			■	■	■	■	■	
HDPE	■	■	■	■	■	■	■	■	■	■	■	
LDPE	■			■		■	■	■	■		■	
MIXED THERMOPLASTICS												
NYLON												
PET	■			■			■	■	■	■	■	
POLYOLEFINS	■	■		■			■	■	■	■	■	
PP												
PS	■	■				■	■	■	■	■		
PVC	■	■	■	■	■	■	■	■	■	■	■	

Appendix A: Plastic Scrap Handlers and Brokers

RECYCLED PLASTICS INDUSTRIES (R.P.I.)

Lee Anderson, President
1820 Industrial Drive
Green Bay, WI 54302

RESIN BROKER
SCRAP HANDLER

PHONE: (414) 468-4545
FAX: (414) 468-4765

R.P.I. buys regrind and buys/sells obsolete or surplus virgin resins. They operate a cleaning or processing system that handles post-consumer scrap and make recycled plastic extrusions for pallets, decking, benches, tables, etc. (and custom applications). Along with processing plastic, R.P.I. designs and engineers equipment to do this work and provides consulting and system design services.

RESINS:	FORM						SOURCES				PREFERRED SHIPMENT	
	WHOLE	GROUND	SHREDDED	BALED	FLAKED	REPROCESSED	PLASTICS PROCESSORS	OTHER MANUFACTURERS	POST COMMERCIAL	POST CONSUMER	TRUCK LOADS	RAIL CAR LOADS
ABS												
ACETALS												
ACRYLIC												
ENGINEERING THERMOPLASTICS												
HDPE		☐	☐	☐	☐		☐		☐	☐	☐	
LDPE												
MIXED THERMOPLASTICS												
NYLON												
PET												
POLYOLEFINS												
PP												
PS												
PVC												

Mixed Plastics Recycling Technology

RIVERSIDE MATERIALS   **RESIN BROKER**

Mary Stachowicz, Business Development Manager
800 South Lawe St.
P.O. Box 815
Appleton, WI 54912   FAX: (414) 749-2130

Riverside Materials buys/sells off-specification or off-grade resins, regrind and obsolete or surplus virgin resins. They have been operating for 10 years and deal in wastepaper as well as plastics.

RESINS:	FORM						SOURCES				PREFERRED SHIPMENT	
	WHOLE	GROUND	SHREDDED	BALED	FLAKED	REPROCESSED	PLASTICS PROCESSORS	OTHER MANUFACTURERS	POST COMMERCIAL	POST CONSUMER	TRUCK LOADS	RAILCAR LOADS
ABS	✓	✓				✓						
ACETALS	✓	✓					✓	✓			✓	
ACRYLIC	✓	✓				✓						
ENGINEERING THERMOPLASTICS	✓	✓	✓	✓	✓	✓	✓	✓			✓	✓
HDPE	✓	✓	✓	✓	✓	✓	✓	✓	✓	✓	✓	✓
LDPE	✓	✓	✓	✓	✓	✓	✓	✓	✓	✓	✓	✓
MIXED THERMOPLASTICS												
NYLON	✓	✓				✓	✓	✓			✓	
PET	✓	✓		✓	✓	✓	✓	✓		✓	✓	
POLYOLEFINS	✓	✓	✓	✓	✓	✓	✓	✓			✓	
PP	✓	✓	✓	✓	✓	✓	✓	✓	✓		✓	✓
PS	✓	✓	✓	✓	✓	✓	✓	✓	✓	✓	✓	✓
PVC	✓	✓	✓	✓	✓	✓	✓	✓	✓	✓	✓	✓

Appendix A: Plastic Scrap Handlers and Brokers 183

WISCONSIN PLASTIC DRAIN TILE CORP.

Gary Fish, President
P.O. Box 8694
Madison, WI 53708

RESIN BROKER
SCRAP HANDLER

PHONE: (608) 884-9437
FAX: (608) 884-6306

Wisconsin Plastic Drain Tile buys/sells regrind and sometimes obsolete or surplus virgin resins. They operate a cleaning or processing system that handles post-consumer scrap and make recycled plastic corrugated plastic tubing. They have been in operation for 15 years.

RESINS:	FORM						SOURCES				PREFERRED SHIPMENT	
	WHOLE	GROUND	SHREDDED	BALED	FLAKED	REPROCESSED	PLASTICS PROCESSORS	OTHER MANUFACTURERS	POST COMMERCIAL	POST CONSUMER	TRUCK LOADS	RAILCAR LOADS
ABS												
ACETALS												
ACRYLIC												
ENGINEERING THERMOPLASTICS												
HDPE	☐	☐	☐	☐	☐	☐	☐	☐	☐	☐		
LDPE												
MIXED THERMOPLASTICS												
NYLON												
PET				☐						☐	☐	
POLYOLEFINS												
PP												
PS												
PVC												

WOODLAND PLASTICS RECYCLING CORP.

RESIN BROKER
SCRAP HANDLER

Gary A. Rice, Vice President of Marketing
2955 Packers Ave.
Madison, WI 53704

PHONE: (608) 241-5690
FAX: (608) 241-8759

Woodland Plastics Recycling buys/sells off-specification or off-grade resins, regrind and obsolete or surplus virgin resins. They perform custom grinding, densifying and conducting. They purchase film and parts scrap, regrind and fiber waste.

RESINS:	FORM						SOURCES				PREFERRED SHIPMENT	
	WHOLE	GROUND	SHREDDED	BALED	FLAKED	REPROCESSED	PLASTICS PROCESSORS	OTHER MANUFACTURERS	POST COMMERCIAL	POST CONSUMER	TRUCK LOADS	RAIL CAR LOADS
ABS	☑	☑	☑	☑	☑	☑	☑	☑	☑		☑	☑
ACETALS												
ACRYLIC												
ENGINEERING THERMOPLASTICS	☑	☑	☑	☑	☑	☑		☑	☑		☑	☑
HDPE	☑	☑	☑	☑	☑	☑		☑	☑		☑	☑
LDPE	☑	☑	☑	☑	☑	☑		☑	☑		☑	☑
MIXED THERMOPLASTICS												
NYLON	☑	☑	☑	☑	☑		☑	☑	☑		☑	☑
PET	☑	☑	☑	☑	☑		☑	☑	☑		☑	☑
POLYOLEFINS	☑	☑	☑	☑	☑		☑	☑	☑		☑	☑
PP	☑	☑	☑	☑	☑		☑	☑	☑		☑	☑
PS	☑	☑	☑	☑	☑		☑	☑	☑		☑	☑
PVC	☑	☑	☑	☑	☑		☑	☑	☑		☑	☑

Appendix B: Sources of Information on Plastics Recycling

B.1 Trade Publications

Plastics News
Bi-weekly newspaper
Crain Communications
1725 Merriam Road
Akron, OH 44313
(216) 836-9180
(800) 992-9970
$20/yr

Plastics Engineering
Monthly magazine
Society of Plastics Engineers Editorial and Business Office
14 Fairfield Drive
Brookfield, CT 06804-0403
(203) 775-0471
$40/yr. (for non-SPE members)

Plastics World
Monthly magazine
P.O. Box 173306
Denver, CO 80217
(800) 662-7776
$74.95/yr

Modern Plastics
Monthly magazine
P.O. Box 602
Hightown, NJ 08520-9955
(800) 257-9402
$39.75/yr

B.2 Plastic Recycling Publications

Resource Recycling
Monthly magazine
Resource Recycling, Inc.
P.O Box 10540
Portland, OR 97210
(800) 227-1424
$42/yr.

Plastics Recycling Update
Monthly newsletter
Resource Recycling, Inc.
P.O Box 10540
Portland, OR 97210
(800) 227-1424
$85/yr.

Recycling Times
Bi-weekly newspaper
National Solid Wastes Management Association
5615 W. Cermak Rd.
Chicago, IL 60650
$95/yr

Reuse/Recycle
Monthly newsletter
Technomic Publishing Co.
851 New Holland Ave
P.O. Box 3535
Lancaster, PA. 17604-3535
$125/yr

Plastic Bottle Reporter
Quarterly newsletter
Society of the Plastics industry
1275 K St. NW, Suite 400
Washington, DC 20005
Free.

B.3 Buyers & Sellers of Waste Plastic

1989 Plastic Recycling Directory
Plastic Bottle Institute
Society of the Plastics Industry
1275 K Street NW, Suite 400
Washington, DC 20005
(210) 371-5319

1990-91 Directory of U.S. & Canadian Scrap Plastic Processors & Buyers / annual
Annual directory
Resource Recycling, Inc.
P.O Box 10540
Portland, OR 97210
(800) 227-1424
$40 (free to subscribers of Plastic Recycling Update)

Appendix B: Sources of Information on Plastics Recycling   187

Directory of Companies Involved in Recycling of Vinyl (PVC) Plastics
The Vinyl Institute
Wayne Interchange Plaza II
155 Route 46 West
Wayne, NJ
(210) 890-9299
Free.

Illinois Recycled Materials Market Directory
ILENR/RR-87/01
Report
Illinois Department of Energy and Natural Resources
Office of Solid Waste and Renewable Resources
325 W. Adams Street, Room 300
Springfield, IL 62704-1893
(800) 252-8955
Free.

Industrial Materials Exchange Service
Illinois EPA
2200 Churchill Road #31
P.O. Box 19276
Springfield, IL 62794-9276
(217) 782-0450
Free.

Wisconsin Companies which Purchase Recycled Plastic
Wisconsin Energy Bureau, Department of Administration
P.O. Box 7868
Madison, WI 53707
(608) 266-8234
Free.

B.4 Plastic Recycling Reports/Proceedings

Guide for Recyclers of Plastic Packaging in Illinois
ILENR/RR-90/10
Illinois Department of Energy and Natural Resources
Office of Solid Waste and Renewable Resources
325 W. Adams Street, Room 300
Springfield, IL 62704-1893
(800) 252-8955
Free.

Methods to Manage and Control Plastic Waste
NTIS PB90-163106
National Technical Information Service
Springfield, VA. 22161
(703) 487-4600
$42.

New Developments in Plastics Recycling - Proceedings of 1989 SPE RETEC Conference
Society of Plastic Engineers
14 Fairfield Drive
Brookfield, CT 06804-0403
(203) 775-0471

Recycling Technology of the 90's - Proceedings of 1990 SPE RETEC Conference
Society of Plastic Engineers
14 Fairfield Drive
Brookfield, CT 06804-0403
(203) 775-0471

Plastics Recycling as a Future Business Opportunity - Proceedings of the 1989 RecyclingPlas IV Conference
Plastic Institute of America at Stevens Institute of Technology, Hoboken, NJ 07030

Plastics Recycling as a Future Business Opportunity - Proceedings of the 1990 RecyclingPlas V Conference
Plastic Institute of America at Stevens Institute of Technology, Hoboken, NJ 07030

Appendix B: Sources of Information on Plastics Recycling 189

B.5 Not-for-Profit Service Organizations Involved in Plastic Waste/Recycling

American Institute of Chemical
Engineers
345 E. 47th Street
New York, NY 10017
(212) 705-7660

Association of New Jersey Recyclers
P.O. Box 625
Absecon, NJ 08201
(609) 641-8292

Association of Plastic Manufacturers in
Europe
Secretariat, 250
Brussels 02-640-2850

Center for Marine Conservation
1725 DeSales St.
Washington, DC 20036
(202) 429-5609

Center for Plastics Recycling Research
Rutgers, Busch Campus
Pistacaway, NJ 08855
(201) 932-3683

Controlled Release Society
16 Nottingham Dr.
Lincolnshire, IL 60069
(708) 940-4277

Council for Solid Waste Solutions
1275 K. St. NW
Washington, DC 20005
(202) 371-5319

Defenders of Wildlife
1244 19th St.
Washington, DC 20036
(202) 659-9510

Degradable Plastics Council
1000 Executive Park
St. Louis, MO 63141-6397
(314) 576-5207

Environment and Plastics Institute of
Canada
1262 Don Mills Rd.
Don Mills, ONT M3B 2W7
(416)449-3444

Environmental Defense Fund
1616 P St. NW
Washington, DC 20036
(202) 387-3500

Environmental Law Institute
1616 P St. NW
Washington, DC 20036
(202) 328-5150

Flexible Packaging Association
1090 Vermont Ave.
Washington, DC 20005
(202) 482-3880

Institute for International Research
13555 Bel Red Rd.
Bellvue, WA. 98009
(800) 468-7644

National Association for Plastic
Container Recovery
5024 Parkway Plaza
Charlotte, NC 28217
(704) 357-3250

National Soft Drink Association
1101 16th St.
Washington, DC 20036
(202) 463-6732

National Solid Wastes Management
Association
1730 Rhode Island Ave.
Washington, DC 20036
(202) 659-4613

Plastic Bag Association
505 White Plains Rd.
Tarrytown, NY 10591
(914) 631-0909

Plastic Bottle Institute
1275 K. St. NW
Washington, DC 20005
(202) 371-5245

B.5 Not-for-Profit Service Organizations Involved in Plastic Waste/Recycling (Cont.)

Plastic Drum Institute
1275 K. St. NW
Washington, DC 20005
(202) 371-5309

Plastic Waste Management Institute
Eukide Bldg.
4-1-13
Tokyo 105
03-437-2251

Plastic Waste Management Institute
Secretariat 250, A
Brussels
026-402850

Plastics Institute of America
Castle Point Rd.
Hoboken, NJ 07030
(201) 420-1606

Plastic Recycling Foundation
1275 K. St. NW
Washington, DC 20005
(202) 371-5200

Polymer Processing Institute
Castle Point Rd.
Hoboken, NJ 07030
(201) 420-5880

Polystyrene Packaging Council
1025 Connecticut Ave
Washington, DC 20036
(202) 822-6424

Rensselaer Polytech Institute
The Pittsburgh Bldg.
Troy, NY 12180-3590
(518) 276-6098

Society of Packaging Professionals
Reston International
Reston, VA 22091
(703) 620-9380

Society of Plastic Engineers
14 Fairfield Dr.
Brookfield, CT 06804
(203) 775-0471

Stevens Institute of Technology
Castle Point Rd.
Hoboken, NJ 07030
(201) 420-5819

U.S. Environmental Protection Agency
Office of Solid Waste
401 M st. SW
Washington, DC 20460
(202) 382-5649

Vinyl Institute
Wayne Interchange
Wayne, NJ 07470
(201) 890-9299

Appendix C: Manufacturers of Plastic Recycling Equipment

SORTING, SEPARATING, CLEANING and DRYING

Air Systems/Aspirators/Dryers
BSB Recycling GmbH
Carter Day
Gala Industries
Kice Industries
Kongskilde Ltd.
Process Control
Sterling Systems

Turnkey Separation Systems
Advanced Recycling Technologies
AKW Apparate & Verfahren GmbH
American Recovery Corp.
Bruce Mooney Associates
Conair Reclaim Technologies
Count Recycling Systems, Inc.
Extrudaids Ltd.
Harris Group
The Heil Co.
Herbold
John Brown Inc., Plastics Recycling Systems
Kloosterman Environmental Equipment
Lindemann Recycling Equipment, Inc.
Lummus Development Corp.
Mayfran International
Miller Manufacturing
Mosley Machinery Co.
New England CRInc.
Procedyne Corp.
Recycled Polymers
Recycling Equipment Manufacturing, Inc.
Reifenhauser-Van Dorn
Reko BV
Secondary Polymers
Sorema
VanDyk Baler Corp./Bollegaf
Wedco
Weiss GmbH
wTe Corp.

Magnetic/Electronic Separators
Bunting Nagnetics
Carpco
Center for Recycling Research
Eriez Magnetics
Industrial Magnetics
Lock International
MSL Pre-Heat AB
Stearns Magnetics

Vibratory Screeners
Kason Corp.
Sweco

Other
Bio Tech Industries-wastewater treatment
International Process Equipment

SIZE REDUCTION

Granulators
Alpine
Altenburger Maschinen
American Pulverizer
Amut
Ball & Jewell
Battenfeld Gloucester
Bauemeister
Blackfriars
Conair
Condux
Constructions Mechaniques Boyard (CMB)
Cumberland/John Brown
J.H. Day/Taylor-Stiles
Erema, c/o Reifenhauser-Van Dorn
Extrudaids
Fabbrica Bondenese Macchine (FMB)
Fitzpatrick
Gloenco
Godding & Dressler
Granutec
Herbold
International Process Equipment
L-R Systems
Mateu & Sole
Moser
Nelmor
Pallmann Pulverizers
Plastmachines
Polymer Machinery
Quickdraft
Ramco Plastics
Rapid Granulator
Reinbold
Sprout Bauer
Tria

Shredders
American Pulverizer
Ball & Jewell
Bepex
Conair
Cumberland/John Brown

Shredders (Continued)
J.H. Day/Taylor-Stiles
Extrudaids
Fitzpatrick
Herbold
Lancelin
Morgardshammar
Nelmor
Pallmann Pulverizers
Quickdraft
Shred-Tech Ltd.
SSI Shredding
Tria

Balers
American Baler
Balemaster
Global Strapping Systems
International Baler
Mosley
Van Dyk

Guillotine Cutters/Bale Openers
J.H. Day/Taylor-Stiles
Fitzpatrick
Herbold
Spadone

Pulverizers
American Pulverizer
Bauermeister
Condux
Godding & Dressler
Herbold
Wedco

Mills
Altenburger Maschinen
Bauermeister
Bepex
Condux
J.H. Day/Taylor-Stiles
International Process Equipment
Pallmann Pulverizers
Wedco

Cryogenic Grinders
Sprout Bauer
Air Products
J.H. Day/Taylor-Stiles
Godding & Dressler
Wedco

Densifiers/Agglomerators
Condux
Cumberland/John Brown
Extrudaids
Fabbrica Bondenese Macchine (FBM)
Fitzpatrick
International Process Equipment
Pallmann Pulverizers
Plastmachines
Weiss GmbH

CONVEYING AND FEEDING

EMI Recycling Systems
Filterless Conveying Systems
IMCS
Kice Industries
Magnificent Machinery/Turtle
Pelletron

REPROCESSING EQUIPMENT

Compounding Extruders
APV Chemical Machinery
John Brown
Instamelt Systems
Wayne Machine & Die
Welding Engineers
Wormser

Filtration Systems
Beringer
Berlyn Corp.
Gneuss
High Technology
PATT Filtration
Welding Engineers

SYSTEMS INTEGRATORS

Separation/Reprocessing Systems
Advanced Recycling Technologies
AKW Apparate & Verfahren GmbH
American Recovery Corp.
APV Chemical Machinery
Browning Ferris Industries
Chambers Development Co.
Conair Reclaim Technologies
Container Recovery Corporation
CRInc.
Dominion Recycling Systems, Inc.

Appendix C: Manufacturers of Plastic Recycling Equipment

Separation/Reprocessing Systems (Continued)
Duraquip, Inc.
Erema, c/o Reifenhauser-Van Dorn
Extrudaids Ltd.
Fairfield County Recycling, Inc.
Future Design
Herbold/Refakt
John Brown Inc., Plastics Recycling Systems
Luumus Development Corp.
M.A. Industries
Magnificent Machinery/Turtle
Materials Recycling Corporation of America
Modern Machinery
New England CRInc.
Ogden Martin Systems
Omni Technical Services
Procedyne Corp.
Recycled Polymers
Reko BV
Raytheon Service Company
REI Distributors
Resource Conservation Services, Inc.
Resource Recovery Systems
RRT-Empire Returns Corporation
Secondary Polymers
Sorema
Waste Management, Inc.
Waste Services Technology
Wedco
Welding Engineers
Weiss GmbH
Wheelabrator Environmental Systems
wTe Corp

Commingled Processing Systems
Advanced Recycling Technologies
Extrudaids

SUPPLIERS OF RECYCLING EQUIPMENT-BY MANUFACTURER

ABB Sprout-Bauer
Sherman Street
Muncy, Pa. 17756
(717)546-1294

Advanced Recycling Technologies
c/o Mid-Atlantic Plastic Systems
Box 507, 320 Chestnut St.
Roselle, N.J. 07203
(201)241-9333

Air Products
7201 Hamilton Blvd.
Allentown, Pa. 18195
(215)481-4520

AKW Apparate & Verfahren GmbH
Georg-Schiffer-Strasse 70
D-8452 Hirschau, West Germany
(49)(09622)183 30

Alpine (Micron Powder Systems)
10 Chatham Rd.
Summit, N.J. 07901
(201)273-6360

Altenburger Maschinen
D-4700 Hamm 1
Vorsterhauserweg 46, West Germany
(49)(02381)4220

American Baler
Hickory Street
Bellevue, Ohio 44811
(419)483-5790

American Pulverizer
5540 W. Park Ave.
St. Louis, Mo. 63110
(314)781-6100

American Recovery Corp.
900 19th St. N.W., Suite 600
Washington, D.C. 20006
(202)457-6613

Amut
Via Cameri, 16-28100
Novara, Italy
(39)(0321)471701

Anderson Plastics
Box 1186
Miami, Okla. 74355
(918)542-7614

APV Chemical Machinery
Polymer Machinery Div.
901 Durham Ave.
Sourth Plainfield, N.J. 07080
(201)561-3700

Automated Recycling Corp.
P.O. Box 13159, Airport Station
Sarasota, Fla. 34278
(813)756-9678

Balemaster
980 Crown Ct.
Crown Point, Ind. 46307
(219)663-4525

Ball & Jewell
5200 W. Clinton Ave.
Milwaukee, Wis. 53223
(414)354-0970

Battenfield Gloucester
Box 900
Gloucester, Mass. 01930-0900
(617)281-1800

Bauermeister
4127 Willow Lake Blvd.
Memphis, Tenn. 38118
(901)363-0921

Bausano Group
Via Bellini 1
1-20095 Cusano Milano, Italy
(39)2-6195221

Bepex
233 N.E. Taft St.
Minneapolis, Minn. 55413
(612)331-4370

Beringer
P.O. Box 485
Marblehead, Mass. 01945
(617)631-6300

Appendix C: Manufacturers of Plastic Recycling Equipment 195

Berlyn Corp.
Bowditch Drive
Worcester, Mass. 01605
(508)852-2233

BioTech Industries
170 Lawlins Park
Wyckoff, N.J. 07481
(201)891-1336

Blackfriars
18-20 Holmethorpe Ave.
Redhill, Surrey RH1 2NL England
(01)(0737)762935

BSB Recycling GmbH
Reuterweg 14, Postfache 101501
D-6000 Frankfurt, West Germany
(49)(069)159-0

Bunting Magnetics
500 S. Spencer Ave., Box 468
Newton, Kans. 67114
(316)284-2020

Carpco
4120 Haynes St.
Jacksonville, Fla. 32206
(904)353-3681

Carter Day
500 73rd Ave. N.E.
Minneapolis, Minn. 55432
(612)571-1000

Center for Recycling Research
Rutgers University
Bldg. 3529, Busch Campus
Piscataway, N.J. 08855
(201)932-3683

Conair Reclaim Technologies
4537 N. Hwy. 29
Newman, Ga. 30265
(404)463-3302

Conair Wortex
400 Haryy S. Truman Pkwy.
Bay City, Mich. 48706
(517)686-6600

Condux
1935 S. Alpine Rd.
Rockford, Ill. 61108
(800)526-6389

Conex of Georgia
173 Kilgore Rd., Ste. 204
Carrollton, Ga. 30117
(404)832-0884

Constructions Mechaniques Boyard
Avenue de la Gare B.P. 17
01910 Bellignat, France
(33)(74)77 91 10

Cumberland/John Brown
Box 6065
Providence, R.I. 02940
(401)728-1600

J.H. Day/Taylor-Stiles
4932 Beech St.
Cincinnati, Ohio 45212
(513)841-3600

EMI, Recycling Systems Div.
427 W. Pike St.
Jackson Center, Ohio 45334
(513)596-6121

Erema
c/o Reifenhauser-Van Dorn
35 Cherry Hill Dr.
Danvers, Mass. 01923
(508)777-4257

Eriez Magnetics
Asbury Road at Airport, Box 10608
Erie, Pa. 16514-0608
(814)833-9881

Extrudaids Ltd.
c/o Action Industries
Box 322, Ellington, Conn. 06029
(203)872-4660

Fabbrica Bondenese Macchine (FBM)
Via Per Zerbinate 29/A
Bondeno, Italy
(0532)896015

Filterless Conveying Systems
R.R. 1
Branchton, Ontario N0B1L0
(519)754-4077

Fitzpatrick
832 Industrial Dr.
Elmhurst, Ill. 60126-1179
(708)530-3333

Foremost Machine Builders
23 Speilman Rd.
Box 644
Fairfield, N.J. 07006
(201)227-0700

Future Design
271 Glidden Rd., Unit 1
Brampton, Ontario L6W1H9
(416)453-9978

Gala Industries
Rt. 2, Box 142
Eagle Rock, Va. 24805
(703)884-2589

Global Strapping Systems
6370 Irwindale Ave.
Azusa, Calif. 91702
(213)404-8111

Gloenco
Box 656
Newport, N.H. 03773
(603)863-1270

Gneuss
2270 Cabopt Blvd. West, Ste. 2271
Langhorn, Pa. 19047
(215)752-5550

Godding & Dressler
Heidestrasse 3 5309
Meckenheim, West Germany
(49)(0 2225)2011

Govoni SpA
via Bondenese, 12
44041 Casumaro, Italy
(30)51-6849105

Granutec
Box 537, Davis Street
East Douglas, Mass. 01516
(508)476-3801

Herbold
36 Maple Ave.
Seekonk, Mass. 02771
(508)761-7120

High Technology Corp.
144 South St.
Hackensack, N.J. 07601
(201)488-0010

Hustler Conveyor Co.
4985 Fyler Ave.
St. Louis, Mo. 63139
(314)352-6000

IMCS
120 W. Washington St.
Zeeland, Mich. 49464
(616)772-1500

Industrial Magnetics
01700 M-75 S., Box 80
Boyne City, Mich. 49712
(616)582-3100

Instamelt Systems
Box 1714
Midland, Tex. 79702
(915)686-5990

International Baler
5400 Rio Grande Ave.
Jacksonville, Fla. 32205
(904)358-3812

International Process Equipment Co.
9300 Rt. 130 N.
Pennsauken, N.J. 08110
(609)665-4007

John Brown Inc., Plastics Recycling Systems
Box 6065
Providence, R.I. 02940
(401)728-1600

Kason Corp.
1301 E. Linden Ave.
Linden, N.J. 07036
(201)486-8140

Appendix C: Manufacturers of Plastic Recycling Equipment

Kice Industries
Box 11388
Wichita, Kans. 67202
(316)267-4281

Kongskilde Ltd.
231 Exeter Rd. E.
Exeter, Ontario N0M-1S3
(519)235-0840

Lancelin
232 Chaussee Jules Ceasar
95250 Beauchamp, France
(33)(1)3413 6943

Lock International
12380 Race Track Rd.
Tampa, Fla. 33626
(813)854-1616

L-R Systems
Box 35
New Lenox, Ill. 60451
(815)485-2155

Lummus Development Corp.
Box 2526
Columbus, Ga. 31902
(404)323-1801

M.A. Industries
303 Dividend Dr.
Peachtree City, Ga. 30269
(404)487-7761

Mateu & Sole
Potosi 9-11 Int. 08030
Barcelona, Spain
(34)(93)345 55 00

Micronyl Wedco
77130 Montereau, France
(33)(1)60 96 13 17

Mid-America Recycling
Box 3312
Des Moines, Iowa 50316
(515) 265-1208

Modern Machinery
3031 Guernsey Rd., Box 423
Beaverton, Mich. 48612
(317)435-9071

Morgardshammar
9800-1 Southern Pine Blvd.
Charlotte, N.C. 28217
(701)522-8024

Moser
Seligenstadter Strasse 83, Postfache 801166
6450 Hanau 8, West Germany
(49)(06181)60051-52

Mosley
Box 1552
Waco, Tex. 76703
(817)779-2491

MSL Pre-Heat AB
Box 104 S-263
22 Hognanas, Sweden
(0)42 31175

Nelmor
Box 328
Rivulet Street
North Uxbridge, Mass. 01538
(508)278-6801

New England CRInc.
74 Salem Rd.
North Billerica, Mass. 01862
(508)667-0096

NRM-Steelastic
1557 Industrial Pkwy.
Akron, Ohio 43310
(216)663-0505

Pallmann Pulverizers
820 Bloomfield Ave.
Clifton, N.J. 07012
(201)471-1450

Partek Corp.
Box 1387
Vancouver, Wash. 98666
(206)695-1777

Patt Filtration
219 S. Main St., Box 1068
Flemington, N.J. 08822
(201)788-8759

Pelletron Corp.
Box 10142
Lancaster, Pa. 17602
(717)293-4008

Plastics Forming Systems
850 E. Industrial Park Dr., Unit 14
Manchester, N.H. 031103
(603)668-7551

Plastmachines
Box 1129
Mableton, Ga. 30059
(404)941-5614

Polymer Machinery
Box 7177
Kensington, Conn. 06037
(203)828-0501

Procedyne Corp.
221 Somerset St.
New Brunswick, N.J. 08901
(201)249-8347

Process Control
Box 47578
Atlanta, Ga. 30340
(404)449-8810

Quickdraft
1525 Perry Dr.
Southwest Canton, Ohio 44708
(216)477-4574

Ramco Plastics
165 Flat River Rd.
Coventry, R.I. 02816
(401)823-0144

Rapid Granulator
5217 28th Ave., Box 5887
Rockford, Ill. 61125
(815)399-4605

Recycled Polymers
32102 Howard St.
Madison Heights, Mich. 48071
(313)583-4279

Refakt
see Herbold

Reifenhauser-Van Dorn
35 Cherry Hill Dr.
Danvers, Mass. 01923
(508)777-4257

Reinbold
Petersbergstrasse 5-5200
Siegburg, West Germany
(49)(02241)38-1104

Reko BV
De Asselenkuil 15, Postbus 191 6190 AD Beek(L)
NL-6161 RD Geleen/L Netherlands
(599)(90)7 60 60

Secondary Polymers
5151 Bellevue Ave.
Detroit, Mich. 48211
(313)922-7000

Shred-Tech Ltd.
Box 119
Gloucester, Mass. 01931
(508)281-6302

Sikoplast/Heinnriich Koch
Box 1816
D-5200 Siegburg, West Germany
(02241)64194

Sorema
Via Provinciale Per Lecco
Lipomo Como ZSITCO
20093 Italy
(39)(031)63 16 37

Spadone
507 Westport Ave.
Norwalk, Conn. 06856
(203)846-1677

Sprout-Bauer
Sherman Street
Muncy, Pa. 17756
(717)546-1294

SSI Shredding
Box 869
Wilsonville, Ore. 99070
(503)682-3633

Appendix C: Manufacturers of Plastic Recycling Equipment

Stearns Magnetics
6001 S. General Ave.
Cudahy, Wis. 53110
(414)769-8000

Sterling Systems
Perrowville Road, Box 219
Forest, Va. 24551-0219
(804)525-4030

Superwood International
Corke Abbey, Bray
County Wicklow, Ireland
(353)1-823835

Sweco
7120 New Buffington
Florence, Ky. 41042
(606)727-5133

Tria
c/o A.C. Hamilton
5820 Kennedy Rd.
Mississauga, Ontario L4Z1T1
(416)890-0055

Turtle Plastics Group
2366 Woodhill Rd.
Cleveland, Ohio 44106
(216) 791-2100

Van Dyk Plastics
234 Fifth Ave.
New York, N.Y. 10001
(212)683-2610

Wayne Machine & Die
100 Furler St.
Totawa, N.J. 07512
(201)256-7374

Wedco
Box 397
Bloomsbury, N.J. 08804
(201)479-4181

Weiss GmbH
Kasseler Strasse 44
Industriegebiet Nord
D-6340 Dillenburg, West Germany
(02771)32795

Welding Engineers
5 Sentry Pkwy. E., Ste. 101
Blue Bell, Pa. 19422
(215)825-6900

Welex Inc.
850 Jolly Rd.
Blue Bell, Pa. 19422
(215)532-8000

Williams & Mettle Co.
3007 Crossview
Houston, Tex. 77063
(713)782-2432

Wormser GmbH
Klosterstrasse 22
6520 Wormms 1, West Germany

wTe Corp.
7 Alfred Circle
Bedford, Mass. 01730
(617)275-6400

Xaloy
Box 1441, Rt. 99
Pulaski, Va. 24301
(703)980-7560

Appendix D: Glossary

ABS (Acrylonitrile-Butadiene-Styrene) A family of thermoplastics based on these three compounds. ABS resins are rigid, hard, tough, and not brittle. This family of plastics is used to produce durable goods products such as appliances and automotive parts.

Acrylic A family of resins formed from methacrylic acid and known for their optical clarity. Widely used in lighting fixtures because they are slow burning or may be made self extinguishing.

Additive A substance added to a basic resin to alter the physical or chemical properties of the resin.

Air Classification The process of passing feed material through or past an air stream at a certain velocity to remove contaminants. The process generally works best with a process stream of dichotomous masses (or densities).

Blow-Molding A method of fabrication in which a parison is forced into the shape of a mold cavity by interval gas pressure.

Coefficient of Thermal Expansion The fractional change in length of a material for a unit change in temperature.

Coextrusion The process of extruding two or more materials through a single die so that the material bonds together at the mating surface.

Colorant A dye or pigment added to a resin to impart color to the plastic.

Compressive Modulus The ratio of compressive stress to compressive strain below the proportional limit. Theoretically equal to the Young's Modulus determined from tensile tests.

Compressive Strength The maximum load sustained by a test specimen in a compressive test divided by the area of the specimen.

Copolymer Typically a polymer of two chemically distinct monomers.

Dewatering Centrifuges and presses are used to remove excess water from plastic particles.

Drying Particles are dried in mechanical and thermal driers to a final moisture content of typically less than 5%.

Elastic Limit The greatest stress which a material is capable of developing without any permanent strain remaining upon complete release of the stress.

Electrostatic Precipitation Primarily aluminum particles from PET beverage bottle caps (and other metals, if present) are passed through an electric field created and collected on plates because of the charge imposed on the particle.

EPM/EPDM (Ethylene Propylene Rubbers) A group of elastomers (rubber-like material) obtained by copolymerization of ethylene and propylene for EPM and a third monomer (diene) for EPDM. Their properties are similar to those of rubber.

EVA (Ethyl-Vinyl Acetate Copolymer) Copolymers of major amounts of ethylene with minor amounts of vinyl acetate that retain many of the properties of polyethylene but have considerably increased flexibility, elongation and impact resistance. EVA is used as a hot melt adhesive for bonding base cups to PET beverage bottles and labels to bottles. EVA is a form of LDPE.

Extrusion The process of forming continuous shapes by forcing molten plastic through a die. Typical shapes are hoses, flat sheets or parisons for blow molding.

Filler A substance (typically inert) added to a plastic compound to reduce its cost or improve physical properties such as hardness, stiffness, impact strength, thermal conductivity or electrical conductivity.

Flake Typically 1/4" or less chips of ground plastic which have been washed, cleaned and dried.

Flexural Modulus The ratio, within the elastic limit, of the applied stress on a test specimen in flexure to the corresponding strain in the outermost fibers of the specimen.

Flexural Strength The maximum stress in the outer fiber at the moment of crack or break. In the case of plastics, this value is usually higher than the straight tensile strength. Also known as modulus of rupture.

Gaylord A container for holding waste plastic, plastic flake or plastic pellets. Often times a gaylord is a cardboard box measuring 34"x43"x38".

Grafted Polymers Polymers may be grafted with a compatibilizer to increase recyclability. A grafted polymer refers to a polymer comprising molecules in which the main backbone chain of atoms has attached to it at various points side chains containing different atoms or groups from those in the main chain. The main chain may be a copolymer or or may be derived from a single monomer. Examples of compatibilizers include methyl methacrylate (MMA)-grafted PE for PE/PVC blending, styrene grafted PE copolymers for PE/PS blends and styrene grafted PVC for PVC/PS blends .

Granulation/Ground Plastic The size reduction of plastic containers to approximate 3/8" or less particle sizes. Granulators have rotating and stationary knives that cut the material. The most efficient models have a tangential feed with blades positioned to produce a scissor-like cut. Capacities range from a few hundred pounds to 5,000 pounds per hour.

HDPE (High Density Polyethylene) Polyethylene plastic having a density typically between 0.940 and 0.960 g/cm^3. While LDPE chains are branched and linked in a random fashion, HDPE chains are linked in longer chains and have fewer side branches. The result is a more rigid material with greater strength, hardness, chemical resistance and a higher melting point than LDPE.

Homopolymer A polymer resulting from the polymerization of a single monomer; a polymer consisting substantially of a single type of repeating unit.

Inhibitor A substance capable of stopping or retarding an undesired chemical reaction. They can be used to prolong storage life or retard degradation by heat and/or light.

Industrial Scrap Plastic material originating from a variety of in-plant operations and which may consist of a single material or a blend of a known composition.

Injection Molding The process of manufacturing with plastic by forcing molten plastic into a mold under pressure.

Izod Impact Strength A measure of impact strength (described in ASTM D256) determined by the difference in energy of a swinging pendulum before and after it breaks a notched specimen held vertically as a cantilever beam. The pendulum is released from a vertical height of two feet, and the vertical height to which it returns after breaking the specimen is used to calculate the energy lost.

LDPE (Low Density Polyethylene) Polyethylene plastic having a density typically between 0.910 and 0.925 g/cm^3. The ethylene molecules are linked in random fashion, with the main chains of the polymer having long and short side branches. The branches prevent the formation of a closely knit pattern, which results in a soft, flexible and tough material.

LLDPE (Linear Low Density Polyethylene) LLDPE is manufactured at much lower pressures and temperatures than LDPE. LLDPE has long molecular chains without the long chain side branches of LDPE, but with the short chain side branches.

Melt Index A single point identification of resin melt viscosity, measured in grams per 10 minute period passing through a specific orifice size at a certain temperature, as dictated by test method ASTM D1238.

Mixed Plastic A mixture of plastics, the components of which may have widely differing properties.

Monomer A compound which typically contains carbon and is of a low molecular weight (compared to the molecular weight of plastics), which can react to form a polymer by combination with itself or with other similar compounds.

Nylon A generic name for a family of resins which have a recurring amide groups (-CO-NH-) as an integral part of the main polymer chain. Nylons are identified by denoting the number of carbon atoms in the polymer chains of each of the constituent compounds which formed the resin. For example, nylon 6,6 refers to the number of carbon atoms in each of the two compounds used to form it.

Off-Spec Plastic Resin that does not meet the manufacturer's specifications.

Parison A hollow tube or other preformed shape of thermoplastic compound which is inflated inside a mold in the blow-molding process.

Pelletizing Cleaned flakes (of a high purity level) are melted and extruded to produce small pellets similar to virgin resin pellets.

PBT (Polybutylene Terephthalate) Similar to PET, but formed using butanediol rather than ethylene glycol (as with PET). PET and PBT are the two thermoplastic polyesters that have the greatest use.

PC (Polycarbonate) PC is characterized by clarity (with optical applications), impact strength and heat resistance. PC is more often used in durable goods production than disposable goods. An easily recognized PC product is 5 gallon water cooler bottles.

PE (Polyethylenes) A family of resins made by polymerizing the gas ethylene (C_2H_4). By varying the catalysts and methods of the polymerization process, its properties can be varied over wide ranges. Some different forms are L/LDPE and HDPE.

PET (Polyethylene Terephthalate) A saturated thermoplastic polyester formed by condensing ethylene glycol and terephthalic acid. It is extremely wear and chemical resistant and dimensionally stable. It also has a low gas permeability in comparison to HDPE, LDPE, PP and PVC which is why it is used so extensively for carbonated beverage bottles.

Phenolics A family of thermosetting resins made by reacting a phenol with an aldehyde. Phenolics are known for good mechanical properties and high resistance to temperature.

Plasticizer A substance or material incorporated into a plastic to increase its flexibility or workability. It may reduce the melt viscosity (or increase the melt index) or lower the resin melting point.

Polyolefins Polymers of relatively simple olefins such as ethylene, propylene and butene. LDPE, HDPE, PP and EVA are polyolefin polymers.

Polyesters A family of resins also known as alkyds. The main polymer backbone is formed through the condensation of polyfunctional alcohols and acids. Polyesters can be saturated (elements or compounds cannot be added to the main backbone) or unsaturated. One of the most important polyester is PET, a saturated polyester.

PP (Polypropylene) A thermoplastic resin made by polymerizing propylene with suitable catalysts. Its density of approximately 0.90 g/cm^3 is among the lowest of all plastics.

Primary Recycling The processing of waste into a product with characteristics similar to those of the original product.

PS (Polystyrene) Polymers of styrene (vinyl benzene). PS is somewhat brittle and is often copolymerized or blended with other materials to obtain desired properties. HIPS (high impact PS) is made by adding rubber or butadiene copolymers. Commonly known PS foams are produced by incorporating a blowing agent during the polymerization process or injecting a volatile liquid into molten PS in an extruder.

PUR (Polyurethanes) A large family of resins based on the reaction of isocyanate with compounds containing a hydroxyl group. PUR can be made into foam or resin, rigid or flexible, thermoset or thermoplastic.

PVC (Polyvinyl Chloride) PVC is produced by polymerization of vinyl chloride monomer with peroxide catalysts. The pure polymer is hard and brittle, but becomes soft and flexible with the addition of plasticizers.

Recycled Plastic Plastic products or parts of a product that have been reground for sale or use to a second party, or plastics composed of post-consumer material or recovered material only (which may or may not have been processed).

Regrind Plastic Plastic products or parts of a product that have been reclaimed by shredding and granulating for use (primarily intended as an in-house term).

Resin A term which is generally used to designate a polymer, a basic material for plastic products. It is somewhat synonymously used with "plastic," but "Resin" (and polymer) most often denotes a polymerized material, while "plastic" refers to a resin which also includes additives such as plasticizers, fire retardants, fillers or other compounds.

Rinsing With all contaminants separated from the plastic containers, clean water is used to rinse off contaminants and any other chemical substance used during washing.

SAN (Styrene-Acrylonitrile Copolymer) A copolymer of about 70% styrene and 30% acrylonitrile, with higher strength, rigidity and chemical resistance over standard PS.

SBS (Styrene-Butadiene-Styrene) A thermoplastic rubber. The styrene and butadiene components are incompatible and form separate phases when mixed which result in rubber-like properties.

Secant Modulus The ratio of total stress to corresponding strain at any specific point on a stress-strain curve.

Secondary Recycling The processing of waste into materials which have characteristics less demanding than those of the original plastic product.

Shredding The size reduction of plastics to 1" to 3" chips. Shredding may be performed prior to granulation to allow for removal of large contaminants such as metal or wood which may damage the granulator.

Sprue In a mold, the sprue is the main feed channel that connects the mold filling orifice to the runners leading to each cavity gate. The term is also used for for the piece of plastic material formed in the channel.

Tensile Strength The maximum tensile stress sustained by a specimen during a tension test. The result is usually expressed in psi, with the area being that of the original specimen at the point of rupture rather than the reduced area after breakage.

Thermoplastic Plastic that can be repeatedly softened by heating and hardened by cooling through a temperature range characteristic of the plastic, and that in the softened state can be shaped by flow into articles by molding or extrusion.

Thermoset Plastic that, after having been cured by heat or other means, is substantially infusible and insoluble. Cross-linking between molecular chains of the polymer prevent thermosets from being melted and resolidified.

Washing Granulated material is washed in a tank with hot or cold water with surfactants (detergents) to loosen and remove contaminants adhering to plastic particles.

Wet Granulation Hot or cold water is added during the granulation process to soften and separate some of the contaminants in the containers.

Yield Strength The stress at which a material exhibits a specified limiting deviation from the proportionality of stress to strain. Unless otherwise specified, the stress will be the stress at the yield point.

Young's Modulus The ratio of tensile stress to tensile strain below the proportional limit.

References

Bond, B. "Recycling Plastics in Akron, Ohio" Proceedings of the Society of Plastic Engineers Regional Technical Conference - Recycling Technology of the 90's. Chicago, IL. 1990.

Bonsignore, P., Jody, B., and Daniels, E. "Separation Techniques for Auto Shredder Residue" Proceedings of the Society of Automotive Engineers International Congress and Exposition. Detroit, MI. 1991.

Briston, J. Plastic Films. Longman Scientific & Technical. Essex, U.K. 3rd Edition. 1989.

Carrier, K. "Reprocessing of Mixed Plastic Scrap" Proceedings of 1989 RecyclingPlas IV Conference. Plastics Institute of America. Hoboken, NJ. 1989.

CNT (Center for Neighborhood Technology). Large Scale, Multi-Resin Plastics Recycling in Chicago. Prepared for Amoco Chemical Company, Chicago, IL. 1990.

Curlee, T. The Economic Feasibility of Recycling - A Case Study of Plastic Wastes. Praeger. New York, NY. 1986.

Dittman, F. "New Developments in the Processing of Recycled Plastics" Proceedings of the Society of Plastic Engineers Regional Technical Conference - Recycling Technology of the 90's. Chicago, IL. 1990.

Franklin Associates. Characterization of Plastic Products in Municipal Solid Waste - Final Report. Prepared for Council for Solid Waste Solutions. Prairie Village, KS. 1990.

Hammer, F. "Production and Marketing of Products from Mixed Plastic Waste" Proceedings of Society of Plastics Engineers RETEC Conference - New Developments in Plastics Recycling. Charlotte, NC. 1989.

Hanesian, D., Merriam, C., Pappas, J., Roche, E., Rankin, S., and Bellinger, M. Post-Consumer Plastic Collection - A Source of New Raw Material From Municipal Solid Waste. Journal of Resource Management and Technology 18(1):35-39, 1990.

Hill, J. Rubbermaid wants Higher Quality Recycled Resin from Plastic Suppliers. Recycling Times 2(12):3, 1990.

Lashinsky, A. NRT Process Sorts PVC from Other Plastics Plastics News 2(37):3, 1990.

Lynch, J., and Nauman, E. "Separation of Commingled Plastics by Selective Dissolution" Proceedings of Society of Plastics Engineers RETEC Conference - New Developments in Plastics Recycling. Charlotte, NC. 1989.

Mack, W. "Turning Mixed Plastic Waste into Specification Products through Advanced Technology" Proceedings of the Society of Plastic Engineers Regional Technical Conference - Recycling Technology of the 90's. Chicago, IL. 1990.

Mackzo, J. An Alternative to Landfills for Mixed Plastic Waste. Plastics Engineering 46(4):51-53, 1990.

Maldas, D., and Kokta, B. Effect of Recycling on the Mechanical Properties of Wood Fiber-Polystyrene Composites, Part I. <u>Polymer Composites</u> 11(2):77-83, 1990.

Miller, E. <u>Plastic Products Design Handbook - Part B: Process and Design for Processes</u>. Marcel Dekker, New York, NY. 1983.

Monks, R. New Reclaim Methods Target PVC. <u>Plastics Technology</u> 36(5):31-32, 1990.

Morrow, D., and Merriam, C. "Recycling - Collection Systems for Plastics in Municipal Solid Wastes - A New Raw Material" <u>Proceedings of Society of Plastics Engineers RETEC Conference - New Developments in Plastics Recycling</u>. Charlotte, NC. 1989.

Morton-Jones, D. <u>Polymer Processing</u>. Chapman and Hall, New York, NY. 1989.

McMurrier, M. Assessing a Polymer's Recyclability. <u>Plastic Recycling Machinery and Equipment Report</u> (Supplement to <u>Plastics Machinery & Equipment and Plastics Compounding</u>). September, 1990.

Mulligan, T. "Commercially Viable Products from Plastic Waste - The Superwood Process" <u>Proceedings of Society of Plastics Engineers RETEC Conference - New Developments in Plastics Recycling</u>. Charlotte, NC. 1989.

Newell, T., and Lewis, M. <u>Informational Report - Development of an Automated Clear/Color Sorting System for Recyclable Containers</u>. Presentation at Conference on Solid Waste Management in the Midwest. Chicago, IL. 1990.

Nosker, T., Renfree, R., and Morrow, D. Recycle Polystyrene, Add Value to Commingled Products. <u>Plastics Engineering</u> 46(2):33-36, 1990.

Phillips, E., Morrow, D., Nosker, T., and Renfree, R. "The Processing and Properties of Recycled Post-Consumer Plastics-Commingled Molding" <u>Proceedings of 1989 RecyclingPlas IV Conference</u>. Plastics Institute of America. Hoboken, NJ. 1989.

Phillips, E. and Alex, A. The Economics of Reclamation of Generic Plastics from Post-Consumer Wastes. Presented at the ACS MARM Conference, Florham Park, NJ. 1990.

Rankin, S., Ed., Frankel, H., Hanesian, D., Merriam, C., Nosker, T. and Roche, E. <u>Plastics Collection and Sorting: Including Plastics in a Multi-Material Recycling Program for Non-Rural Single Family Homes</u>. Center for Plastics Recycling Research, Rutgers, The State University of New Jersey, Pistacaway, NJ. 1988.

Rankin, S. Recycling Plastic in Municipal Solid Wastes. <u>Journal of Resource Management and Technology</u> 17(3):143-148, 1989.

Rankin, S. <u>Plastics Recycling Processes</u>. Rutgers Center for Plastics Recycling Research. Pistacaway, NJ. 1990.

Rennie, C. "The Pitfalls and Promises of Plastics Recycling" <u>1990 Proceedings of the RecyclingPlas V Conference</u>. Plastics Institute of America. Hoboken, NJ. 1990.

Renfree, R., Nosker, T., Rankin, S., Kasternakis, T., and Phillips, E. "Physical Characteristics and Properties of Profile Extrusions Produced for Post-Consumer Commingled Plastic Waste" <u>1989 Proceedings of Society of Plastic Engineers Annual Technical Conference (ANTEC)</u>. Brookfield, CT. 1989.

Summers, J., Mikofalvy, B., and Little, S. Use of X-Ray Fluorescence for Sorting Vinyl from Other Packaging Materials in Municipal Solid Waste. <u>Journal of Vinyl Technology</u> 12(3):161-164, 1990a.

Summers, J., Mikofalvy, B., Wooton, G., and Sell, W. Recycling Vinyl Packaging Materials from the City of Akron Municipal Wastes. <u>Journal of Vinyl Technology</u> 12(3):154-160, 1990b.

Tirrell, D. "Copolymerization" pp. 201-204 in the <u>Concise Encyclopedia of Polymer Science and Engineering</u>. J. Kroschwitz, Editor. John Wiley & Sons. New York, NY. 1990.

U.S. EPA Office of Solid Waste. <u>Methods to Manage and Control Plastic Wastes - Report to Congress</u>. Publication Number EPA/530-SW-89-051. Washington, D.C. 1990a.

U.S. Environmental Protection Agency (EPA) Office of Solid Waste. <u>Characterization of Municipal Solid Waste in the United States: 1990 Update</u>. Publication Number EPA/530-SW-90-042A. Washington, D.C. 1990b.

Vane, L., and Rodriguez, F. "Selective Dissolution: Multi-Solvent, Low Pressure Solution Process for Resource Recovery from Mixed Post-Consumer Plastics" <u>Proceedings of the Society of Plastic Engineers Regional Technical Conference - Recycling Technology of the 90's</u>. Chicago, IL. 1990.

Vernyl, B. Research with Thermosets Getting Renewed Attention. <u>Plastics News</u> 2(29):15, September 17, 1990.

Wielgolinski, L. "A Family of Functionalized Acrylic Polymers with Unique Solubility Properties for Recycling Applications" <u>Proceedings of Society of Plastics Engineers RETEC Conference - New Developments in Plastics Recycling</u>. Charlotte, NC. 1989.

Yam, K., Gogoi, B., Lai, C., and Selke, S. Composites from Compounding Wood Fibers with Recycled High Density Polyethylene. <u>Polymer Engineering and Science</u> 30(11): 693-699, 1990.

Other Noyes Publications

HANDBOOK OF POLLUTION CONTROL PROCESSES

Edited by

Robert Noyes

This handbook presents a comprehensive and thorough overview of state-of-the-art technology for pollution control processes. It will be of interest to those engineers, consultants, educators, architects, planners, government officials, industry executives, attorneys, students and others concerned with solving environmental problems.

The pollution control processes are organized into chapters by broad **problem areas**; and appropriate technology for decontamination, destruction, isolation, etc. for each problem area is presented. Since many of these technologies are useful for more than one problem area, a specific technology may be included in more than one chapter, modified to suit the specific considerations involved.

The pollution control processes described are those that are actively in use today, as well as those innovative and emerging processes that have good future potential. An important feature of the book is that **advantages** and **disadvantages** of many processes are cited. Also, in many cases, **regulatory-driven trends** are discussed, which will impact the technology used in the future.

Where pertinent, regulations are discussed that relate to the technology under consideration. Regulations are continually evolving, frequently requiring modified or new treatment technologies. This should be borne in mind by those pursuing solutions to environmental problems.

Innovative and emerging technologies are also discussed; it is important to consider these new processes carefully, due to increasingly tighter regulatory restrictions, and possibly lower costs. For some pollutants specific treatment methods may be required; however for other pollutants, specific treatment levels must be obtained.

CONTENTS

1. REGULATORY OVERVIEW
2. INORGANIC AIR EMISSIONS
3. VOLATILE ORGANIC COMPOUND EMISSIONS
4. MUNICIPAL SOLID WASTE INCINERATION
5. HAZARDOUS WASTE INCINERATION
6. INDOOR AIR QUALITY CONTROL
7. DUST COLLECTION
8. INDUSTRIAL LIQUID WASTE STREAMS
9. METAL AND CYANIDE BEARING WASTE STREAMS
10. RADIOACTIVE WASTE MANAGEMENT
11. MEDICAL WASTE HANDLING AND DISPOSAL
12. HAZARDOUS CHEMICAL SPILL CLEANUP
13. REMEDIATION OF HAZARDOUS WASTE SITES
14. HAZARDOUS WASTE LANDFILLS
15. IN SITU TREATMENT OF HAZARDOUS WASTE SITES
16. GROUNDWATER REMEDIATION
17. DRINKING WATER TREATMENT
18. PUBLICLY OWNED TREATMENT WORKS
19. MUNICIPAL SOLID WASTE LANDFILLS
20. BARRIERS TO NEW TECHNOLOGIES
21. COSTS

INDEX

In summary, a vast number of pollution control processes and process systems are discussed.

7" x 10"   **750 pages**

Other Noyes Publications

INDUSTRIAL SYNTHETIC RESINS HANDBOOK
Second Edition

Edited by

Ernest W. Flick

This Second Edition of the *Industrial Synthetic Resins Handbook* is a collection of technical data for about 3,000 resins and related products. It is the result of information received from 57 manufacturers and distributors of these products. The data represent selections from manufacturers' descriptions made at no cost to, nor influence from, the makers or distributors of these materials.

It is believed that all of the products listed are currently available, which will be of interest to readers concerned with raw material discontinuances. The book will be of value to technical and managerial personnel involved in the manufacture of final products made from these resins, as well as to suppliers of basic raw materials.

The information provided for each product includes, as available:
Resins Category
Company Name
Trade Name
Product Number
Product Description
Properties/Characteristics

Also included in the book are a list of **Suppliers' Addresses** and a **Trade Name Index**.

The resins and related products are divided into the following 19 sections, with the number of products covered in parentheses:

1. **ABS and SAN Resins (42)**
2. **Acetal Resins (81)**
3. **Acrylics (201)**
4. **Alkyds (434)**
5. **Curing Reaction Resins (62)**
6. **Epoxy Systems and Related Products (289)**
7. **High Solids Resins (186)**
8. **Nylons (130)**
9. **Phenolics (81)**
10. **Polycarbonates (32)**
11. **Polyesters (219)**
12. **Polyethylenes (208)**
13. **Polypropylenes (101)**
14. **Polystyrenes (77)**
15. **Polyurethanes (140)**
16. **Rosins, Rosin Esters and Hydrocarbon Resins (219)**
17. **Silicones (29)**
18. **Vinyls (111)**
19. **Miscellaneous (350)**

ISBN 0-8155-1287-2 (1991)   6" x 9"   1155 **pages**

Other Noyes Publications

HANDBOOK OF TOXIC AND HAZARDOUS CHEMICALS AND CARCINOGENS
1991 — Third Edition

by Marshall Sittig
Princeton University (retired)

"... an important addition to the literature on toxic chemicals.... Marshall Sittig has made it easier for us to understand how to control the handling of toxic chemicals and the disposal of toxic wastes." —from the Foreword by *Bill Bradley, U.S. Senator, New Jersey.*

The *Third Edition* of this Handbook is a completely new, much-expanded version of this widely-accepted classic. It presents chemical, health and safety information on 1300 toxic and hazardous chemicals and carcinogens (up from 800 in the previous edition). The book will aid responsible decision making by chemical manufacturers, safety equipment producers, toxicologists, industrial safety engineers, waste disposal operators, health care professionals, and the many others who may have contact with or interest in these chemicals due to their own or third party exposure. The book is a must for industrial, medical and legal libraries.

Included in the Third Edition are substances which are:
    Designated by **EPA** as "**Hazardous Substances**", "**Hazardous Wastes**" under **RCRA**, "**Priority Toxic Pollutants**", "**Extremely Hazardous Substances**" under **SARA** (particularly the top 100), or "**Toxic Chemicals**";
    Identified as **carcinogens** by the **U.S. National Toxicology Program** and the **IARC**;
    Identified in **UN/DOT** shipping regulations;
    Reviewed in **NIOSH Information Profiles** and **EPA "CHIPS"** documents;
    Set forth by **OSHA**;
    Whose allowable concentrations in air and other safety considerations have been considered by **NIOSH**:
    Whose allowable concentrations in workplace air are adopted or proposed by **ACGIH**, **Deutsche Forschungsgemeinschaft**, or the **Health and Safety Executive** (U.K.);
    Whose allowable concentrations in workplace air, residential air and water have been set by the **USSR-UNEP/IRPTC** project.

In addition the book covers:
    Most of the chemicals in the **ILO** *Encyclopedia of Occupational Health and Safety*, the **United Nations'** *IRPTC Legal File*, or the journal *Dangerous Properties of Industrial Materials Report*, and
    Many chemicals cited in "right-to-know" legislation of one or more of the 50 states in the U.S.A.

Essentially the book attempts to answer these questions about a compound (to the extent information is available): (1) What is it? (2) Where is it encountered? (3) How much can one tolerate? (4) How does one measure it? (5) What are its harmful effects? (6) How does one protect against it? (7) How does one handle it and protect against mishaps?

The chemicals are presented alphabetically and each is classified as a "carcinogen," "hazardous substance," "hazardous waste," and/or a "priority toxic pollutant"—as defined by the various federal agencies. Data and citations are furnished, to the extent currently available, on any or all of these important categories:

Chemical Description	Points of Attack
Code Numbers	Medical Surveillance
Synonyms	First Aid
Potential Exposure	Personal Protective Methods
Incompatibilities	Respirator Selection
Permissible Exposure Limits in Air	Storage
Determination in Air	Shipping
Permissible Concentration in Water	Spill Handling
Determination in Water	Disposal Method Suggested
Routes of Entry	References
Harmful Effects and Symptoms	

ISBN 0-8155-1286-4 (1991)   2 volumes   7" x 10"   1685 pages

Other Noyes Publications

FORMULATING PLASTICS AND ELASTOMERS BY COMPUTER

by

Ralph D. Hermansen
Formula Research Services Co.

This unique book combines expert knowledge of plastic and elastomer formulating processes with computer programming to create novel software tools for the formulator. Pascal language is used to build powerful data structures and to write clear, understandable code.

Computer programs which facilitate various formulating steps can make formulating easier, faster, better, and less expensive. The book discusses how such programs can be written and includes actual source code for the sample programs.

This book was written for people with an interest in the formulation of plastics and elastomers *and* in applying computer programming to that technical area. The same challenge and sense of accomplishment that makes formulating enjoyable also applies to programming.

The Pascal programs in the book require Turbo Pascal Version 4.0, or higher. Further, the programs were written using Turbo Pascal on an IBM compatible computer. Be advised that there are minor differences between Turbo Pascal for IBM and MacIntosh computers. If a MacIntosh is used, some modification will be necessary.

The programs provided in the book are complete and may be input by the user, or a disk containing the programs can be purchased (see box).

CONTENTS

1. The Formulation Task
2. Using Computers for Formulating
3. Review of the Pascal Programming Language
4. Writing Simple Application Programs
5. Making Programs User-Friendly
6. Compound Packaging Program Development
7. Writing a Filled Formulation Program
8. Plastisol Formulation Program Development
9. Epoxy Stoichiometry Program Development
10. Polyurethane Stoichiometry Program Development
11. Adding Databases to Programs
12. Passing Formulations to Procedures
13. Looking Forward
14. List of Programs

COMPANION SOFTWARE PACKAGE

The Pascal source code listing and compiled programs for *Formulating Plastics and Elastomers by Computer* are all included on 3½" or 5¼" disks. For IBM compatible users, the disks will save many hours of typing. Compiled EXE files of the programs can be immediately run and examined. Included on the disks are programs for:

- Reviewing Pascal Programming Principles
- Formulation Costing (one including a database)
- User-Friendly Interfacing
- Developing a Compound Kit
- Simultaneously Determining Property Changes of a Formulation as Filler Is Added
- User-Requested Custom Vinyl Plastisol Formulations
- Calculating Proper Weight Ratios of Epoxy or Urethane Reactants
- Holding, Storing or Utilizing Formulations via Datafiles

Learn from these valuable programs and adapt these new principles to your own applications.

ISBN 0-8155-1276-7 (disk) (1991)
$150 (prepaid, non-refundable)

ISBN 0-8155-1275-9 (book) (1991)   6" x 9"   278 pages

Other Noyes Publications

PLASTIC WASTES
Management, Control, Recycling, and Disposal

by

U.S. Environmental Protection Agency

T. Randall Curlee, Sujit Das
Oak Ridge National Laboratory

Pollution Technology Review No. 201

Plastic wastes in the municipal solid waste (MSW) stream and in the industrial sector are discussed in this book. Quantities of plastic wastes generated, characterization of the wastes, their environmental impact, and management of the waste stream are described. Also covered is the effect of improper disposal of plastic wastes on the marine environment.

As plastic use, and particularly plastic packaging, have grown more prevalent, so have concerns about plastic wastes disposal. Discarded plastics, besides being highly visible, are a rapidly increasing percentage of solid waste in landfills. These problems have made plastic wastes a major focus in the management of solid waste.

The book has two parts. Part I, emphasizing plastic wastes in the MSW stream, provides a technological review of the plastics industry, production and consumption statistics, definitions of major end use markets, and disposal paths for plastics. It examines management issues and environmental concerns on landfilling and incineration, the two primary MSW disposal methods currently in use. Plastic litter and its impact are covered. Source reduction and recycling are considered, and recommended actions for industry and other concerned groups are identified.

Part II addresses issues related to plastics recycling in the industrial sector: manufacturing and post-consumer plastic waste projections, the estimated energy content of plastic wastes, the costs of available recycling processes, institutional changes that promote additional recycling, legislative and regulatory trends, the potential quantities of plastics that could be diverted from the municipal waste stream and recycled in the industrial sector, and the perspectives of current firms in the plastics recycling business.

Several scenarios are presented which assume specific technical, economic, institutional, and regulatory conditions. A **condensed table of contents** is given below.

I. WASTES IN THE MSW STREAM
1. Introduction
2. Production, Use, and Disposal of Plastics and Plastic Products
3. Impacts of Plastic Debris on the Marine Environment
4. Impacts of Post-Consumer Plastics Waste on the Management of MSW
5. Options to Reduce the Impacts of Post-Consumer Plastics Wastes
6. Objectives and Action Items

Appendix A—Statutory and Regulatory Authorities Available to EPA and Other Federal Agencies

Appendix B—State and Local Recycling Efforts

Appendix C—Substitutes for Lead- and Cadmium-Containing Additives for Plastics

II. RECYCLING IN THE INDUSTRIAL SECTOR
7. Introduction
8. Plastic Waste Projections
9. The Energy Content of Plastic Wastes
10. The Cost of Recycling Versus Disposal
11. An Update of Recent Institutional Changes That May Promote Recycling
12. Recent Legislative and Regulatory Actions
13. The Potential for Divertable Plastic Waste
14. Plastics Recycling from the Current Industry's Perspective
15. Conclusions

Appendix—Plastic Waste Projections: Methodology and Assumptions

Other Noyes Publications

CHEMICAL ADDITIVES FOR THE PLASTICS INDUSTRY
Properties, Applications, Toxicologies

by

Radian Corporation

This book describes in detail the chemical additives used in the plastics industry. It analyzes the chemicals used as additives in polymer manufacturing and the processing of plastics, environmental releases of these chemicals, and possible occupational exposures to them.

The development and growth of the use of plastics since World War II has been phenomenal. The technology for the manufacture of polymers and plastic products has expanded into the most important of the chemical-based industries, producing new products for new uses at a remarkable rate. Plastics additives have played a critical and complex role in this growth. The purpose of this book is to put additives used in the manufacture of plastic products in perspective for environmental and health impact analysis.

The plastics additives are presented as major functional groups of chemicals, which are further subdivided into chemically-, functionally-, or physically-similar chemicals. An overview of each major functional group includes the properties and applications of the subclasses, their environmental impacts, and possible occupational exposures. Common worker exposure practices for each functional group are also presented.

A series of appendices details the physical and chemical properties and polymer application of each chemical within the functional groups; the industrial, commercial, and consumer uses and consumption volumes for each chemical; and data on toxicological and worker exposure concerns for each chemical.

CONTENTS

1. INDUSTRY DESCRIPTION
2. ANTIOXIDANTS
3. ANTISTATIC AGENTS
4. BLOWING AGENTS AND OTHER ADDITIVES FOR FOAMED PLASTICS
5. CATALYSTS FOR THERMOPLASTIC POLYMERIZATION
6. COLORANTS
7. COUPLING AGENTS
8. CURING AGENTS AND CATALYSTS FOR THERMOSETTING RESINS
9. FILLERS AND REINFORCERS FOR PLASTICS
10. FLAME RETARDANTS
11. FREE RADICAL INITIATORS AND RELATED COMPOUNDS
12. HEAT STABILIZERS
13. LUBRICANTS AND OTHER PROCESSING AIDS
14. PLASTICIZERS
15. PRESERVATIVES
16. SOLUTION MODIFIERS AND OTHER POLYMERIZATION AIDS FOR PLASTICS
17. ULTRAVIOLET STABILIZERS

REFERENCES

Appendix A—Physical and Chemical Characteristics with Polymer Application for Plastics Additives

Appendix B—Consumption and Other Uses for Plastics Additives

Appendix C—Toxicological and Worker Exposure Concerns for Plastics Additives

ISBN 0-8155-1114-0 (1987)

884 pages

ASHEVILLE-BUNCOMBE
TECHNICAL COMMUNITY COLLEGE

3 3312 00044 8043

TD 798 .H44 1992

Hegberg, Bruce A.

Mixed plastics recycling technology

DISCARDED

JUN 16 2025